CONTROL POLICIES FOR SPECIFIC WATER POLLUTANTS

ORGANISATION FOR ECONOMIC CO-OPERATION AND DEVELOPMENT

The Organisation for Economic Co-operation and Development (OECD) was set up under a Convention signed in Paris on 14th December 1960, which provides that the OECD shall promote policies designed:
- to achieve the highest sustainable economic growth and employment and a rising standard of living in Member countries, while maintaining financial stability, and thus to contribute to the development of the world economy;
- to contribute to sound economic expansion in Member as well as non-member countries in the process of economic development;
- to contribute to the expansion of world trade on a multilateral, non-discriminatory basis in accordance with international obligations.

The Members of OECD are Australia, Austria, Belgium, Canada, Denmark, Finland, France, the Federal Republic of Germany, Greece, Iceland, Ireland, Italy, Japan, Luxembourg, the Netherlands, New Zealand, Norway, Portugal, Spain, Sweden, Switzerland, Turkey, the United Kingdom and the United States.

Publié en français sous le titre :

POLITIQUES DE LUTTE CONTRE
LES POLLUANTS SPÉCIFIQUES DES EAUX

.˙.

The number of pollutants likely to be found regularly in rivers of industrial countries is growing continually and has now reached several thousand. Though part of the increase may be explained by new analytical techniques which can identify a larger range of substances, industrial development continues to add to the number of pollutants released to the water environment. Many of these pollutants have been shown to be hazardous for man and wildlife: they may be acutely toxic or have long-term chronic effects, even at low concentrations. They include a large number of organic compounds, and a number of inorganic substances - heavy metals, for instance. Contamination of groundwaters, less significant in the past, is now also of concern, since traditionally safe drinking water supplies are also being contaminated and may soon become unsafe. Particular contaminants in this case are nitrates (from agricultural fertilizers) and organic solvents (of industrial origin).

Responding to serious concern in Member countries, the OECD Council recommended in 1974 that intensive efforts be made to control specific water pollutants. With this objective, the Environment Committee launched an integrated programme which was carried out by its Water Management Policy Group and completed in 1981. This report describes the results of the programme, including the identification of pollutants, their origin and pathways, and the possible impacts on human health, in particular through drinking water supplies. Practical control measures are assessed.

Also available

POLITIQUE ET INSTRUMENTS DE GESTION DE L'EAU. Série «Documents» (novembre 1977)
(97 77 03 2) ISBN 92-64-21646-4   170 pages   £3.90   US$8.00   F32.00
English text: Water Management Policies and Instruments, out of print – available on microfiches (2):   £1.20   US$2.70   F12.00

WATER MANAGEMENT IN INDUSTRIALISED RIVER BASINS. "Document" Series (May 1980)
(97 80 04 1) ISBN 92-64-12063-7   163 pages   £4.40   US$10.00   F40.00

EUTROPHICATION OF WATERS. MONITORING, ASSESSMENT AND CONTROL (April 1982)
(97 82 03 1) ISBN 92-64-12298-2   154 pages   £5.20   US$11.50   F52.00

Prices charged at the OECD Publications Office.

THE OECD CATALOGUE OF PUBLICATIONS and supplements will be sent free of charge on request addressed either to OECD Publications Office,
2, rue André-Pascal, 75775 PARIS CEDEX 16, or to the OECD Sales Agent in your country.

TABLE OF CONTENTS

Part I

SUMMARY

Part II

ASSESSMENT, CATEGORISATION AND HAZARD RATING OF SPECIFIC WATER POLLUTANTS

Part III

ORIGIN, OCCURRENCE AND CONTROL OF SPECIFIC POLLUTANTS IN POTABLE WATER

Part IV

CONTROL OF ORGANOCHLORINATED COMPOUNDS IN DRINKING WATER
(An example of a practical control strategy)

Part I

SUMMARY

I.1. INTRODUCTION

Water pollution by defined chemical substances has a long history; classical examples in the past are phenols and pyridines in effluents from coke oven plants.

Since the mid-1960s, however, people have become aware that some substances - for instance mercury derivatives, DDT and PCBs - can persist in the environment, enter and become enriched in food chains and reach toxic levels in certain organisms. This recognition of toxicity has been a driving force behind the development towards stricter and better environmental management. Furthermore, studies sparked off by the recognition of these effects led to the realisation that contaminants from human activities are ubiquitous in the environment, and that many of these substances are potentially hazardous to man's environment. There is undoubted concern because:

- information on the type and quantity of substances discharged to the environment is incomplete;
- with a few notable exceptions, the information on environmentally significant properties, particularly stability, affinity for living organisms and short- and long-term toxicity is incomplete and scarce;
- a number of these compounds are not adequately removed by present treatment of waste water or potable water, and for other substances there is insufficient information;
- available control instruments - legal, administrative and others, and their implementation - do not always cover all situations where the discharge of specific pollutants should be controlled.

The specific (water) pollutants(*) concept was used to emphasize the difference, on the one hand, between

---

*) Specific (water) pollutant was defined, for the purpose of this study, as "a substance which is mainly introduced into the environment by human activity and which, under given conditions, lowers the quality and value of a water resource, particularly by toxic and nuisance effects on human beings or aquatic life". Specific pollutants have also been called "trace pollutants", "micropollutants", and "refractory pollutants". Other organisations, such as the EEC, have used different definitions.

identifiable chemical substances, and on the other hand classical aggregate or general parameters such as BOD(*), COD(*) and suspended solids. Some pollutants, for instance mercury and cadmium compounds and phenols, are defined chemical species. Other groups, such as cationic surface active agents, consist of closely related compounds which are sufficiently similar for the whole group to be considered as an entity in water management. A third group consists of a complex mixture of a large number of compounds - such as a textile process effluent - for which a substance-by-substance approach is extremely difficult.

Because of the complexity of the problem, it was not judged possible to carry out a complete assessment on specific pollutants control. For example, not all the characteristics of non-point sources were identified and although methods for analysis of costs of control of some specific pollutants have been improved, a complete analysis could not be made. The main objectives were to review the technologies and strategies for pollutant control, and to propose tools for coherent assessment and priority setting.

Specific pollutants have been found in large numbers in all types of natural waters ranging from obviously polluted industrial rivers to deep groundwaters which were thought to be protected from pollution. Estimates based on the total number of chemicals used in industry (and their probable break-down metabolites or by-products) suggest that the number of pollutants released in the environment is enormous. Most of these specific pollutants present in natural waters have not yet been identified. Industrial development is continuously increasing the number of specific pollutants in the environment. Besides, identification of these pollutants is limited by sensitivity of analytical techniques. Surveys in OECD Member countries have identified several hundred specific pollutants.

Many pollutants have already been shown to be hazardous to man and to aquatic life, and others, which are not acutely toxic, have been shown by laboratory tests or epidemiological studies to have long-term chronic effects. Finally, even nuisance effects may prove very costly, when, for example, supplementary stages are necessary in the preparation of potable water.

The capability of traditional waste water treatment plants to reduce, to acceptable levels, pollution as measured by classical parameters, is now recognised as insufficient. Furthermore, this study illustrated the limited ability of these plants to remove specific

*) Biological Oxygen Demand; Chemical Oxygen Demand.

11

pollutants. For example, recent comparisons of BOD, COD
and TOC data have shown that treatment plants do not
remove certain types of organic compounds. The potential
toxic, carcinogenic and nuisance effects of many pollutants are therefore sufficiently serious to encourage
more rigorous examination of sources of pollution and
measures by water authorities to combat them.

I.2. PRINCIPLES FOR EVALUATING SPECIFIC POLLUTANTS

Recognising that monitoring of specific pollutants and quantification of the problem is difficult, and acknowledging that action is needed to prevent deterioration of water quality, the Water Management Group proposed a framework and guidelines for gathering quantitative and qualitative data on specific pollutants (see Part II for further details).

Lists of hazardous chemicals - "black lists" - based on existing information and experience, although useful, are not exhaustive and cannot possibly relate to all practical conditions. The Group therefore decided to try to develop a general, preventive and predictive approach better to protect potable water supplies and aquatic ecosystems, and which should meet water authorities' requirements for an evaluation system in day-to-day use.

The system takes into account the information required for decisions on control of pollution by widely occurring chemical substances and preparations. Data on, for instance, chemical and physical properties; potential effects on different kinds of organisms; use patterns; and removability from waters should be compiled and analysed before the appropriate level of control can be finally established. It is often necessary to make administrative decisions based on present knowledge before a full analysis of the situation can be made. Part II gives a basis for a hierarchical system for identifying substances which may endanger water quality. The system allows new information to be incorporated quickly. It is primarily intended as a checklist for experts working with regulatory authorities with the ultimate responsibility for evaluating hazards. The rational and consistent information structure is to aid overview and comparison and prevent inadvertent omission of important factors.

It was also recognised that those responsible for implementing water management policies would be helped if the hazard potential of specified water pollutants could be expressed by a simple formula which would assist in their ranking. A number of existing or proposed systems for rating hazards were first examined, but the Group noted that these were mainly intended for particular situations - such as the GESAMP profiles for the seaborne transport of chemicals.

A hazard rating system for general use in water management needs to cover a number of complex, frequently dissimilar situations, in which substances may be intentionally or unintentionally, partially or completely dispersed in waters. A choice had to be made between developing either an elaborate and complex rating designed to cover a maximum number of situations or a simple rating for alerting or early warning. While recognising the possibility of misuse of a simple rating system, the Group proposed an easily read formula which summarises particularly relevant data and alerts the water authority that the production, use or disposal of a particular pollutant may interfere with the quality of the water resources and that special protective measures may be required.

The system is called the THP rating, which stands for Toxic or harmful effects in the aquatic environment (T), Hazard to human health (H), and biochemical Persistence (P). The rating has three numbers related (on a logarithmic scale) to:

i) the lowest concentration at which the pollutant has an adverse effect on aquatic ecology;

ii) the lethal dose for oral intake by man (LD_{50});

iii) the maximum half-life of the specific pollutant in the aquatic environment.

The THP rating is easily understood, since the greater the value of the component numbers the greater is the need to consider regulatory measures. It is stressed that the rating has to be used with discrimination and it should not be used for purposes for which it was not designed. (See Part II for further details of the Hazard rating system.)

Lowest concentration having an adverse effect in the aquatic environment (g/l)	Toxicity T	Quantity giving rise to hazard to human life by direct oral intake LD_{50} (g/kg)	H	Estimated maximum half-life of pollutant in aquatic environment (days)	Persistence P
1	0	1	0	1	0
10^{-1}	1	10^{-1}	1	10	1
10^{-2}	2	10^{-2}	2	100	2
10^{-3}	3	10^{-3}	3	1 000	3
10^{-4}	4	10^{-4}	4		
10^{-5}	5	10^{-5}	5		
10^{-6}	6	10^{-6}	6		
10^{-7}	7	10^{-7}	7		
10					

Suspected or established carcinogens are designated by adding a 'c' after the H value.
If a pollutant tends to be bioaccumulated a '+' is shown.

I.3. SOURCES OF SPECIFIC POLLUTANTS

Sources of pollutants may differ widely in technical and environmental significance as well as in ease of control. The principal types are point sources and non-point (diffuse) sources but there are some overlapping and borderline situations.

A. POINT SOURCES

Until the present, point sources have been the major category. They include:

Industrial sources:

- extraction and primary processing of minerals and fuels;
- industrial production of materials and chemicals;
- utilisation and transformation of materials and chemicals by industry;

Municipal sources:

- domestic, commercial, service and small industries.

a) Industrial sources

i) Extraction and processing of minerals and fuels

Several forms of mineral extraction give rise to effluents or spills which, without preventive measures are likely to impair water quality. Decisive factors are the character and richness of the particular mineral and the local technical, geological and hydrogeographical conditions. Typical pollutants from mineral extraction are toxic metals and sulphuric acid from weathered sulphidic ores, sulphuric acid and certain organic chemicals from pyritic coals and alkaline earths, metal sulphates and chlorides from rocksalt deposits. Mainly for economic reasons, OECD Member countries have tended to close the smaller operations, leaving large-scale and long-term activities. Extraction of a particular mineral is often confined to a few sites.

Plants for processing and upgrading ores and minerals are frequently located near the extraction site. Effluents from these plants generally contain the same pollutants as those from mining as well as, some toxic, auxiliary chemicals used in enrichment processes such as flotation.

The extraction of certain minerals generates large amounts of spoil. As considerable land areas are usually required for disposal, spoil deposits become diffuse rather than point sources. Drainage water from inadequately managed spoil deposits from some minerals, particularly sulphidic ores, may be acidic and rich in dissolved toxic metals. Abandoned extraction sites and spoil deposits may generate polluted drainage waters for a long period, and particular control problems arise when legal responsibility for an abandoned site is difficult to establish.

The extraction of petroleum resources and the processing of crude oil involve the risk of discharges of noxious organic substances such as aromatic hydrocarbons, phenols and sulphur-containing compounds. Severe risks arise from accidental spills at well sites or during transportation.

ii) Industrial production of materials and chemicals

As the primary origin of a major number of specific water pollutants, the chemical industry is an obvious starting point for study of man-made chemicals which include:

- heavy chemicals, including fertilizers, primary plastics, rubber etc.;
- other bulk materials such as non-ferrous metals and pulp;
- speciality chemicals;
- chemical preparations.

Economic production of heavy chemicals and other bulk materials requires large installed capacities on relatively few sites. The large flow of raw materials into, and of products out of the sites demands transportation facilities. The situation is typified by petrochemicals; the same type of approach can be used for heavy chemicals and with modifications for other bulk materials such as non-ferrous metals and cellulosic pulps.

Heavy chemicals are manufactured mainly by continuous and high-yielding processes. High capital intensity and tailored single-purpose process technology generally make for stable production profiles and changes have so

17

far been mostly step-by-step addition of new units and replacements of obsolete ones. Furthermore, as feedstocks and chemical reactions used are usually well defined, the chemical character of the majority of the non-recovered materials is relatively easy to establish. This means that effective monitoring may be exercised. Since the total turnover of materials exceeds one million tonnes per annum at many sites, the load of specific pollutants generated will be considerable and thus special control measures will be needed.

The manufacture of <u>speciality chemicals</u> such as pharmaceuticals, fine chemicals, dyes, paints and glues is more difficult to survey as the number of processes and product types is very large. Equipment is frequently multi-purpose to allow flexible production programmes, and processes tend to be discontinuous and done batch-wise. Because product yields range from moderate to high, the number of non-utilisable by-products may be considerable. Demands for raw materials, process water and transportation facilities are generally limited, which make location requirements less strict.

Production of speciality chemicals ranges from 100 tonnes to 50,000 tonnes per product per site per year. For individual fine chemicals, annual production may only be a few tonnes at one site, but several hundred products may be manufactured.

The quantity and composition of effluents from production of speciality chemicals vary widely both over time and from plant to plant. Small plants generally give only elementary treatment to effluents which are then discharged to public sewers; in contrast, at large factories on-site effluent treatment is often followed by discharge to surface waters. The risk of hazardous by-products, which might be ignored (i.e. Seveso), should not be neglected.

iii) <u>Utilisation and Transformation of Materials and Chemicals by Industry</u>

Transformation of raw materials and use of chemicals as reactants, frequently in a dispersive way, are a main source of pollutants. Effluents are likely to contain, as pollutants, part of the materials and reactants used (solvents, detergents, dyes, toxic metals, salts etc.) and also a large number of ill-defined by-products. Lack of characterisation of these by-products contributes to the severity of the problem. There is a great variability between one industry and another which makes a branch by branch approach necessary.

The combination of chemicals by mechanical mixing is the main feature of chemical preparation manufacture, which only occasionally includes chemical reactions. Raw materials are usually supplied from off-site, but on-site production of major ingredients occurs occasionally. High utilisation of raw materials is usually achieved. Water-borne pollutants are usually discharged when equipment has to be cleaned and thus discharges are irregular rather than continuous. Peak overloading of either external or internal treatment facilities can occur, particularly at weekends when equipment is cleaned.

b) <u>Municipal Sources</u>

After industries, the major source of specific pollutants is effluent from municipalities. An increasing part of these effluents is now treated through municipal waste water treatment plants, although the variability is considerable within OECD countries (probably less than 5 per cent to over 80 per cent) and on average about 40 per cent to 50 per cent have some type of treatment. Moreover, the real effectiveness of elimination of pollutants in the treatment process may be unsatisfactory not only for "traditional" parameters (BOD, suspended solids) which are not in fact difficult to remove, but especially for specific pollutants which pass through the treatment without being eliminated. This portion, which is mainly non-biodegradable organic material and inorganic compounds, is the most undesirable part of the pollutant load.

Together with the purely domestic discharges, municipal sewers also receive effluents from commercial and service sectors. These sectors generate characteristic pollutants, for example: detergent constituents (phosphates, borates and complexants) in effluents from laundries; disinfectants from hospitals; emulsifiers, cleaning agents and petroleum products from garages and car service stations; and solvents from printing works. In addition to the pollutants from reasonably defined sources, ubiquitous sources give pollutants such as: plasticisers (particularly dioctylphthalate), fluorescent whitening agents, contraceptives and their degradation products, and metallic compounds (copper, lead and zinc) in soluble form following corrosion.

Pollutants from small and medium-sized industries are likely to vary considerably and although the individual sources are generally small, the combined load can be appreciable - especially when a particular urban zone or river basin has many such industrial units.

B. DIFFUSE SOURCES

Diffuse sources are those where pollutants are dispersed in waters under conditions where there is little or no possibility for direct control. The overall importance of these sources of pollutants can be seen since they are at least as large as the pollutant load from point sources in some river basins, and thus offset the efforts being made to control conventional point sources.

Raw surface and groundwaters vary considerably in their "background" composition reflecting the local geology/geochemistry. Waters in Europe are on average lower in arsenic, molybdenum and vanadium and higher in iron and strontium than waters in North America. Substances naturally occurring in water rarely pose problems for the preparation of potable water but salt and humic acids may be exceptions.

Modern intensive agriculture and forestry may contribute significantly to pollutant levels through:

- spreading of biocides for control of insects, weeds and crop diseases;
- application of chemical fertilisers (especially nitrogen and phosphorus compounds);
- release of organic pollution and nutrients from manuring, intensive animal breeding, ensilage of crops;
- increase in natural nitrate run-off in ground or surface waters following tillage of grassland, ground disturbance and deforestation.

Unless carefully controlled, disposal sites for domestic refuse may become important sources of pollutants because the drainage water can contain a wide spectrum of organic and inorganic compounds and metals. Disposal of mixed industrial and domestic refuse increases the hazard if seepage takes place.

Run-off from urban and industrial areas, harbours, roads and airports generally carries a high pollutant load. Typical constituents are: lead, zinc, asbestos and hydrocarbons from vehicles; zinc, cadmium and copper from corroded metal structures; glycols from de-icing of aircraft and salt from snow removal. In high traffic urban areas, rain storms which occur after long dry periods carry large amounts of pollutants into the sewer system. These large volumes of storm waters may be discharged directly to rivers, or can flood over the holding tanks and thus pass with their untreated pollutants into the water environment. In other situations, the treatment plant becomes overloaded and the efficiency of pollutant removal falls drastically, again resulting in environmental pollution.

Many airborne pollutants may ultimately cause water pollution, through rainfall and dry deposition. Fossil fuel combustion (domestic heating, motor vehicles, power plants) is a major source of sulphur oxides, toxic and radio-active metals, tars etc. Many industrial processes such as steelworks and metal smelters are also significant contributors. Fertilizer and aluminium industries emit fluorides and the petrochemical industry emits numerous organic compounds. These pollutant effects range from nuisance through to severe damage to human health and ecology.

Because of their unpredictable timing and location, accidental releases of pollutants have to be considered with diffuse sources. The most common type of accident occurs during transport and transfer of chemicals by water, rail, road and pipeline. Accidents during storage of liquid chemicals and during emergencies at chemical factories are also considerable.

I.4. POLLUTION CONTROL IN INDUSTRIAL BRANCHES

(Pilot Case Studies)

In order to identify and analyse factors to be considered in the design of programmes for controlling specific pollutants from industry, three pilot studies were done and detailed reports of these are available from OECD. The industrial branches were chosen to illustrate different organisational, technical, economic and pollutant characteristics. Two branches - metal-plating and textile finishing - are characterised by use and consumption of significant quantities of a wide variety of chemicals and chemical preparations; the third - petrochemicals and its downstream manufacturing processes - is the root source of most of the organic chemicals used in industry and households.

A. METAL-PLATING

The metal-plating industry is heterogeneous in the size of its units and the types and uses of its products. The number of plants is large - about one per eight thousand inhabitants in Member countries. Three main types of metal-plating plants are:

- establishments operating as an integrated part of a complex for the manufacture of composite products, for instance electrical equipment (i.e. internal).
- plants used on an integrated basis for finishing mass produced articles such as: automobiles (i.e. external).
- specialised enterprises, often quite small workshops, doing sub-contract work, (i.e. external).

A primary purpose of metal-plating, i.e. the electrolytic coating of a carrier metal (usually steel) with thin layers of other metals, is to improve the durability of goods by increasing their resistance to corrosion and/or wear. Accordingly, electro-platers, whether integrated or on sub-contract, generally have to conform to strict quality requirements. This close link with their customers (both internal and external) is an important aspect of their trade. Furthermore, metal-plating is complementary to a number of industries - such as electrical, engineering and automobiles - and therefore cannot be replaced by importation.

The following industry characteristics are particularly relevant for water management:

- the main components in effluents, generally well known chemicals, do not change but their concentrations may fluctuate widely. Nevertheless, proprietary formulations of unknown composition are frequently used;
- most metal-plating processes are only semi-continuous;
- workshops are numerous and widely scattered;
- the acute toxicity of the process chemicals (cadmium, chromium, nickel and copper, frequently with cyanide in addition) hastened the early development of effluent treatment technology;
- metal-plating is still largely characterised by "craft" rather than by technology; hence, personnel are likely to respond best to simply operated pollution control equipment;
- the auxiliary chemicals used in metal-plating processes may adversely affect effluent treatment, also their composition is frequently unknown to the user or there may be complexants;
- many installations are too small to allow detailed effluent monitoring;
- sometimes, adequate pollution control cannot be achieved without extensive replacement of production plant;
- recent advances in metal-plating technology, especially automation, combined with more efficient utilisation of water and process chemicals offer scope for further improvements of effluent quality;
- sub-contract metal-plating is of key importance, particularly to many engineering industries, and pollution control programmes should, therefore, provide sub-contractors with adequate assistance and time for compliance with control requirements;
- technical research and development is sometimes less important than administrative surveillance and assistance;
- because of predominance of small firms in the metal-plating industry, centralised waste treatment schemes are particularly relevant.

B. TEXTILE FINISHING

Dyeing and finishing of yarns and fabrics, whether done internally or on a sub-contract basis, play a quality and value-improving role in the whole textile industry comparable to the role of metal-plating in certain metal processing industries. However, the dyeing and finishing sector is considerably more diverse and difficult to survey because of:

- a wider range of raw materials and products dif- fering in type, technical properties and quality;
- a larger number of unit processes, each with its own individual requirements for process chemicals;
- a greater complexity of unit processes;
- the greater variety of process chemicals used.

An ample water supply and closeness to centres of production or importation of raw fibres were decisive factors in location of the textile industry and in many Member countries it is still predominantly located in specific regions. The spread of the industry in the post-war period was essentially linked to the birth and rapid growth of man-made fibre manufacture.

Apart from the number and variety of colouring materials, the textile dyeing and finishing industry uses many widely differing auxiliary chemicals, the majority of which are water-borne and not recovered. Unit processes are largely discontinuous and water con- sumption is relatively high. The following additional features are relevant to water management:

- the textile industry is linked with fashion, thus a type of fabric or a particular dye or shade of dye may be needed for part of a season only and then abandoned indefinitely;
- dyeing and finishing plants operate frequently on very short-term programmes; orders for a few hundred metres of fabric for delivery within a week are not uncommon, particularly in the high-quality sector;
- variable product profiles and dependence of many of the employed unit processes on complex com- binations of auxiliary chemicals result in efflu- ents which vary widely over time and from site to site. Data on purchased chemicals may pro- vide a basis for a quantitative estimation of discharges of pollutants;
- some years ago, hydrophilic natural fibres - such as wool and cotton - were processed with natural and biodegradable products. More recent use of lipophilic man-made fibres together with stable synthetic chemicals, has caused increased difficulties in effluent treatment and control;
- dyeing and finishing processes are compatible only with limited recycling of water and process chemicals at present;
- the combination of complexity and variability makes the characterisation of textile effluents, and the design of waste water treatment equip- ment difficult;

24

- considerable research and development is still
needed on effluent control.

C. PETROCHEMICALS

This industry, illustrative of water management
problems in heavy process industries, grew in the late
1940s out of the petroleum refining and heavy organic
chemical industries. Early petrochemical complexes were
located close to, or actually integrated with, either
refineries or organic chemical factories. Later, however, complexes have been constructed independently
of these "parent" industries and petrochemicals is now
a recognisable industrial sector. The key unit of a
petrochemical site is the cracker which converts a
specified feedstock, e.g. naphtha or gas oil, to an
optimal mixture of simple reactive unsaturated hydrocarbons together with minor amounts of hydrogen and
methane. Among the products, ethylene, propylene and
benzene have the greatest economic potential. The ratio
between the primary products from a given cracker is
fixed within narrow limits by engineering design factors.
The other major route to petrochemicals is through the
conversion of natural gas.

Installed capacities of individual crackers started
from ca. 20,000 tonnes and increased rapidly before
levelling out at ca. 450,000 tonnes around 1970. The
feedstock turnover of modern petrochemical sites exceeds
1 million tonnes per annum. Sites are relatively few
(about 1 for every 5-10 million inhabitants) and are
mainly located at the coast or on navigable inland
waterways.

In addition to the cracker, a petrochemical site
has satellite plants in which primary products are further processed to bulk polymers, solvents and raw materials for other industries. Thus, although the production, and hence pollution, profile of a site may
initially give an impression of great complexity, it
can be rationally analysed in a way parallel to the
textile and electroplating industries. The earlier
systematic and fairly rigid hierarchy of unit processes
may be described as a "product tree" whose branches have
generally two to five products.

Petrochemical processes are high-yielding and mostly
continuous. Many have inherent characteristics which
permit products of side reactions to be recovered. Feedstocks are low in non-hydrocarbon matter of the type that
results in noxious by-products, for instance, sulphur
compounds. Although the industry produces relatively

moderate quantities of pollutants per unit of production, the very large-scale production of each plant means that the total load is generally considerable and efficient pollution control facilities will therefore always be necessary. In addition, the following features relevant to water management may be noted:

- untreated effluents from the individual process units of a site are generally fairly constant in pollution compounds apart from variations attributable to normal operational irregularities;

- a large quantity of organic pollutants from petro-chemical processes are either simple, low molecular weight compounds, water-immiscible hydrocarbons or solid polymers; consequently, rigorous application of traditional, physical, chemical and biological unit processes is frequently sufficient for achieving adequate effluent quality for this category of substances. Nevertheless, more complex, persistent and hazardous chemicals are also produced which may have a serious polluting effect unless special treatment is applied;
- particular treatment problems arise with the manufacture of petrochemicals containing chlorine or nitrogen (as well as carbon and hydrogen) because of the persistence and toxicity of the by-products;
- within a site, there may be a number of options for rational choice of pollution control points;
- petrochemical technology is subject to continuing development resulting in, among other things, improvements in the utilisation of raw materials and in the recovery of products and by-products leading to decreased primary discharge of pollutants. However, the introduction of new, less polluting technology is generally restricted to new installations or installations subject to major renovation.

As a major supplier of raw materials, the petro-chemical industry is closely tied to other sectors of the organic chemical industry. However, the problems of water pollution by chemical substances in the broad organic chemical industry are more complicated than those related to petrochemical sites, and problems of hazardous chemicals in effluents from manufacture of organic chemicals in general deserve a separate in-depth study.

D. COMPARATIVE DISCUSSION OF POLLUTION CONTROL IN THE THREE PILOT BRANCHES

From the regulatory point of view, traditional discharge-oriented controls - normally based on technological requirements - can be applied in all three types of industry. The use of product oriented legislation to strengthen pollution control should also be considered.

Pollution control in the three branches also differs from the viewpoint of the administrative support needed. At one end of the scale is electroplating which places a heavy demand for advice and supervision, at the other end is the petrochemicals industry which is, in general, competent enough to organise and manage its pollution control work by itself, provided authorities establish precise objectives, examine submitted discharge data and check that effluent requirements are indeed met. In both electroplating and textile finishing, craftsmanship remains an important feature, and trial and error is still used to test the performance of process chemicals, their environmental properties generally being left to the suppliers. Advice is required by all industries at the time of setting up a plant. Such advice must come from regulatory authorities concerning effluent standards and types of technology available. Authorities, industrial trade associations, and independent consultants will all have a function in advising companies on normal process problems. For small and medium-sized industrial plants special assistance may be required.

The systematic diminution of total volumetric discharges of effluents is a major factor in both the reduction of pollutant emissions and water resource conservation. This can be achieved by the systematic introduction of specific in-plant measures and modifications. This approach has great potential in electroplating as well as in petrochemicals. Its application to textile finishing will, in many cases, require further effort.

The metal-plating industry often works as a separate or integrated unit within large, industrial frameworks (i.e. automobile) which tie it closely to national markets and protect it relatively well against direct international competition. On the other hand, the textile industry is markedly consumer oriented and extremely sensitive to the evolution of the international market. This has made it prone to competition. Under this competitive pressure the textile industry of several OECD Member countries has significantly contracted in recent years, thus creating severe problems for employment and supply of textiles for strategic or similar purposes. In relation to the investment in pollution control, the textile finishing industry exemplifies a sensitive

sector with a lower than average profitability, which
is not expected to change in the near future. To some
extent this feature of low profitability is also found
in the metal-plating industry.

The extensive use of auxiliary chemical preparations,
habitually marketed under non-descriptive trade names is
an important factor in many industries, particularly in
textile finishing and metal-plating. The responsibility
for evaluation of the health and environmental hazards of
the different constituents of these preparations primarily rests with their manufacturers. The awareness of
the chemical industry of its responsibility for product
safety and environmental respect is now growing but this
does not, however, completely solve the problem. Firstly,
it may require a number of years to satisfy the most
urgent need for environmental toxicological data.
Secondly, many manufactured chemicals are not marketed
for immediate industrial and domestic consumption but for
use as ingredients of chemical preparations to be used
in industry and in the domestic, agricultural and commercial sectors.

Primary producers of chemical products have no practical control to prevent their products being misused
and dispersed of in indiscriminate ways by the secondary
users. Implementation of product oriented legislation
is urgently required to provide the means to control this
kind of problem.

Studies analogous to the three carried out would re-
veal that each branch (for instance the tanning, pulp
and paper industries) or sub-section of a branch (such
as in the chemical industry, the chlorine, fertiliser or
pharmaceutical industries) has a particular profile for
the conditions of generation and thus the means of controlling specific pollutants. Within a branch-to-branch
approach, a "multi-barrier" concept of pollution control
should be applied in each industrial plant in order to
better prevent and control all kinds of pollution spills
due to accidents and mismanagement.

I.5. EFFECTIVENESS OF POLLUTANT REMOVAL IN CURRENT TREATMENT PROCESSES

Parallel with the pilot studies on industrial branches were those on the processes for treatment of municipal waste water and for preparation of potable water. The main thrust of these studies was to evaluate the efficiency of the various treatment processes and sequences in removing specific pollutants.

A. WASTE WATER TREATMENT

Conventional waste water treatment involves sequences of unit processes; the combinations adopted are a function of the composition of the raw effluent, and the desired quality level for the treated effluent. The classification of the different treatment processes (such as preliminary, primary, secondary, tertiary) is rather arbitrary and may lead to confusion, unless the types of process are specified. In practice (secondary) biological treatment is the key element of municipal treatment plants in most OECD countries; it may be an activated sludge process, or biological filtration.

Physico-chemical processes are less common. They include, coagulation by metallic salts or polyelectrolytes, precipitation with lime at high pH and activated carbon adsorption. They may be useful for treatment of waste waters where industrial effluents are an important component since they are not affected by substances toxic to biological processes and are particularly effective in removing heavy metals. These processes may also be used in plants with a high variation of flow, or where there is insufficient space to build larger installations, and moreover, when the treatment requirement is only limited to suspended solids and a part of soluble BOD. In practice, best results will be obtained with a combination of physico-chemical processes and biological treatment. The removal of nutrients, such as phosphorus compounds, responsible for eutrophication of waters, is a typical case of the use of physico-chemical processes; these are frequently associated with classical (biological) treatment. Phosphorous is generally precipitated efficiently with lime or metallic sa ts. Nitrogen removal is much more difficult and may be achieved by various (biological, chemical and physical) processes.

Conventional sequences of treatment processes used in municipal plants will effectively remove the major part of the biodegradable pollutants as well as some non-biodegradable substances by adsorption and settlement. However, the uneliminated load may still contain most of the original content of (i) inorganic and (ii) of those organic specific pollutants which are difficult to degrade. It is among these residual substances (known sometimes as "biorefractories") that the most hazardous pollutants often occur but the additional treatment methods are generally uneconomic which means that at-source elimination is essential. Because the removal rate of a substance is a function of its concentration (i) traces of even biodegradable substances inevitably remain in solution in treated waters and (ii) the effectiveness of pollutant removal is, in general, much higher on concentrated effluents.

There is considerable scope for more efficient operation of existing municipal sewage plants and thus reduction of water pollution from these sources. A number of technical improvement measures could be taken, such as increased residence time, addition of physico-chemical steps, and monitoring of effluent, but often the fundamental problems stem from managerial, administrative and financial inadequacies. Basic problems include neglect and lack of interest by municipal authorities in the treatment plant (often considered as an unproductive investment); poorly trained personnel and lack of operating funds; and continuing mismanagement leading to poor pollution abatement. Fundamental issues are:

- a scheme of regular financing to ensure proper operation throughout the life of the plant, with the necessary provisions formally planned at the initial investment stage and guaranteed. For instance, an appropriate charge levied at municipal level from all users in proportion to discharge and abstractions, might guarantee continuous financing where necessary;
- adequate management of plants requires operators with suitable technical qualification. Regular training programmes leading to professional qualification should become compulsory for all operators. Moreover, the operation and inspection of treatment plants should progressively become the responsibility of a specially trained corps of inspectors and operators. As a first step to this permanent arrangement, regular inspections should be carried out at all plants.

B. POTABLE WATER PREPARATION

(For greater detail see Parts III and IV)

Although the quality of drinking water is still satisfactory in most areas of OECD countries, a net deterioration has been noted over the last two decades in the quality of raw waters used; this has also led to a lowering in the chemical quality of the treated waters because a large number of micropollutants are poorly removed by conventional treatment. This deterioration, more often observed in regions of rapid urban and industrial growth, results from the following:

a) Water demand has considerably increased in urban areas during this period because of increased population; increased consumption per capita; increased demands from (industrial, agricultural) activities unduly supplied with drinking water.

b) In order to meet the ever-increasing demand the existing generally good quality, traditional sources have often been supplemented by waters of low quality from "easy-to-get-at" rivers in urban areas. Hence, there has been a net deterioration in the quality of raw water sources.

c) Other contributory factors are:

- long-term reservation of high quality waters for potable use has not taken place;
- proper reallocation and transfer of good quality water resources, hitherto used for non-demanding industrial or agricultural purposes, has not been carried out;
- the trend of supplying an ever-increasing range of non-domestic uses from the potable water network;
- an overestimate of the capabilities of water treatment, and some economic reasons have also led to preferential use of local raw waters, whatever their quality, rather than that of safer waters transported from a distance.

d) Finally, large-scale treatment of these low-quality surface waters by accelerated processes, and use of high doses of chemicals, has frequently given unsatisfactory taste and odour and led to review of the human health hazards. Systematic analysis of water, before and after treatment, has shown that a large number of specific pollutants pass through the treatment. Furthermore, by-products are formed during treatment, including organohalogens, which present a health hazard.

31

Medical research and epidemiological surveys currently being carried out (or already published) suggest some correlations between sources of consumed water and mortality or cancer occurrence. It is too early, and it is not the object of this report to judge this issue. Nevertheless, consumption of water containing even relatively low levels of certain hazardous contaminants may, because of cumulative effects, lead to a significant build-up over a period of years. A dangerous threshold might then be reached after a certain time by the population drawing its water from a limited number of sources over long periods. When there are sufficient presumptions of risk, even long term, for such a population, corrective action is required.

Part III of this report points out that there is clearly no single solution to the problem, and all measures which can contribute to the <u>improvement</u> of <u>raw waters</u>, as well as the improvement of <u>treatment</u> and <u>distribution</u> of potable water will be positive.

Utilisation of better raw waters is the best and the safest approach:

- Progressive improvement of poor quality resources presently used, by stricter pollution control, increase of low flows etc. Nevertheless, in many densely urbanised and industrialised river basins, the sources of contamination may be so extensive that it is unlikely that the water quality of river will recover sufficiently within an acceptable time. Thus it would be advisable to abstract better resources elsewhere.
- An inventory should be made of all surface and underground resources within a reasonable distance, and their quality should be assessed. Available resources of good quality could be used immediately. Where good quality resources are used for activities not requiring such high quality, legal and administrative action should be taken to transfer this water to the drinking water supply - the first priority in the hierarchy of water use.
- Where sufficient resources of good quality cannot be found within a relatively short distance, transportation over longer distances (an "aqueduct" policy) should be considered.

Possible improvements in potable water treatment, may include:

i) an increase in the number of unit processes used to broaden the range of pollutants removed. For instance, this may involve physico-chemical treatments (often operated at a high rate). The outlook for various techniques, such as ion exchange, appears interesting. Apart from the purely technical problems, various other modern technologies, such

32

as reverse osmosis or electrodialysis, are also handicapped by their energy requirements;

ii) use of more comprehensive biological processes (operated at a lower rate) which combine biological, chemical and physical effects such as slow biological (sand or carbon filtration or soil filtration). These soft "environmental" techniques are now favoured over severe physico-chemical processes with potential negative effects (break-point chlorination for instance);

iii) better use of good, proven, classical processes and optimisation of their sequence in order to apply wise management techniques complementary to the treatment itself, for example: better monitoring of raw and finished waters, better location and operation of water uptake, installation of larger storage for raw and finished water - in a word, better "husbandry".

The distribution networks themselves, directly or indirectly, affect the quality of drinking water. Poor maintenance is frequent - losses of 50 per cent are common in certain regions; and the "dirtiness" of piping may lead to excess chlorination of water before distribution, which creates other problems. Asbestos cements, lead pipes, cadmium soldering, have also been identified as potential hazards in distribution.

The use of certain reagents, such as oxidants (e.g. chlorine and ozone), in water treatment has been extensively discussed. These issues have been the subject of Part IV of this report and will not be expanded on here. Nevertheless, the opinion of many experts can be summarised as follows. The extensive use of oxidants, especially at the early stages of treatment, may lead to the formation of significant levels of potentially hazardous by-products (such as organochlorines) by reaction with organic compounds present in raw waters. This practice should thus be discouraged. Disinfection is the essential function of oxidants and should be carried out using a moderate dosage at the final stage of treatment when the organic pollutants have been minimised. The other secondary uses of oxidants in pre-treatment and treatment itself (such as the control of algae or fixed organisms, removal of ammonia by break-point chlorination etc.) are often responsible for the major formation of organohalogens and should be replaced by the available alternative processes.

A general feeling by the experts was a certain modesty in relation to what is currently possible, and the necessity not to overestimate the potential of new technologies. Within reasonable economic limits, it is unrealistic to expect that high quality drinking water, from both taste and health viewpoints, can be obtained from low quality raw waters, at least at present.

I.6. POSSIBLE STAGES IN CONTROLLING SPECIFIC POLLUTANTS

A basic conclusion drawn from the study is that, in practice, there is no single control procedure which can provide a safe barrier to the spread of pollutants, and thus efficient environmental protection.

The great diversity of specific pollutant sources - point or non-point, regular or accidental - implies that controls be applied through a certain number of "barriers" established at the most relevant points of the chain of product generation → use → disposal.

In practice, control may not be feasible or rational at all possible points simultaneously, and it will be necessary to choose the best combination. Such a policy may be best described as a multi-barrier approach. The conceptual scheme for establishing pollution barriers appears from the pilot studies to follow a logical pattern which is summarised in Table I.

A. PRE-PRODUCTION AND PRE-MARKETING CONTROL AND LIMITATIONS OF USE

Pre-production control is the first barrier which can be raised against the spread of pollutants; it has already been used in a number of Member countries. There are a series of levels of this type of control. The first level, used in the development of drugs, pharmaceuticals and biocides, can detect undesirable - or potentially undesirable - environmental effects before commercial scale production starts. This can allow regulatory authorities to control or ban manufacture of the compounds in question. The second level concerns the regulations which non-manufacturing countries may have against the importation of substances. Here regulatory agencies can restrict the entry of products thought to be environmentally undesirable. Finally, the general or particular uses of some materials may be subject to scrutiny by regulatory agencies; and policies of substitution or strict conditions for use can be imposed on the potential users. Further discussion of these issues and the policy questions surrounding them can be found in reports of the OECD Chemicals Group and will not be further dealt with here.

Prevention of pollution at source is by far the most effective and safest means of control. This can be carried out by different strategies: for example, by banning undesirable processes and products and by replacement with less polluting ones; by the use of closed systems including recycling; by the early segregation of industrial effluents and their treatment by specific processes. Furthermore, early prevention and control procedures can diminish the risk of accidental spills. In effect, the later the stage of control the less effective it is likely to be, due to wider dispersion of the contaminants. In principle, measures to limit the dispersal of pollutants should be applied at the point where these substances stop serving their intended purpose. However, in practice, this basic principle may not always be achievable.

Table I.1

MAIN POSSIBLE "BARRIERS" AND STAGES OF CONTROL

EARLY CONTROL OF PRODUCTS, MATERIALS AND TECHNOLOGIES

I. Pre-production and Pre-marketing Control

- before production, importation and sale of products in general;
- leading to possible banning or modification and restriction.

II. Strict Limitations and Specifications concerning Conditions of Use

- for industry, agriculture or general public.

CONTROL OF INDUSTRIAL EFFLUENTS

I. Control of Pollutants at the factory

In general, two possible levels of treatment exist at the factory:

- effluent purified to such a level that it can be discharged directly to surface waters, in conformity with regulations;
- or, only pre-treatment before transfer to municipal treatment plant.

1. At the stage of the Industrial Process itself

This approach is based on optimisation of the manufacturing technology for minimal waste release (for instance "best practicable technology"):

a) At new plants: design and engineering specifications of the process (close cycles, dry processes, new approaches);
b) At existing plants: modification or adjustment of processes, reduction of water use, choice of "cleaner" reactants and materials; and in general stricter management and "husbandry".

2. Following the Industrial Processing

a) Selective treatment of the different effluent streams allowing rational recycling of water, materials and reactants, and a closer control of pollutants.
b) Treatment of Mixed Effluent Streams

II. Control of Effluents outside the Factory

1. Treatment at municipal plant of non-toxic raw effluents or effluents pretreated at the factory to remove toxicity.
2. Treatment at industrial treatment plants, used jointly by several local industries.
3. Treatment at central facilities (detoxification centres) of concentrated effluents, sludges and slurries.

CONTROL OF MUNICIPAL EFFLUENTS

Municipal plants may receive several types of effluent: domestic and commercial; untreated from some types of industries; pretreated from industry; urban run-off. Different levels of treatment may thus be applied:

1. Normal treatment (generally biological) for domestic effluents.
2. Supplementary treatment stages when certain types of effluents (industrial) are present, or for nutrient removal (phosphorus precipitation).
3. Special treatment for urban run-off and storm waters (for instance lagooning).

LAST STAGE OF CONTROL OF POTABLE WATER PLANTS

- Despite the previous "barriers" many pollutants reach the natural waters abstracted for drinking purposes (residual pollutant load unremoved by waste water treatment; uncontrolled discharges; accidents; non-point sources).
- The number of treatment units (Biological/Physical/Chemical) and their combination must be adjusted to the degree of pollution in the raw waters. The reagents for treatment and disinfection must be carefully used in order not to generate new pollutants.
- For its limited rate of removal for specific pollutants, potable water treatment should only be regarded as a last resort security barrier.

B. CONTROL OF INDUSTRIAL EFFLUENTS DURING THE PROCESS ITSELF

The pilot studies on industrial branches have considered a certain number of possible approaches to control pollutants during industrial manufacture; it has been concluded that a pollutant should be removed from waste water as close to the point of its appearance as possible. This generally means that a pollutant can be recovered in a concentrated form in purpose-built equipment. Such recovery appears to be frequently the most economic solution to the problem since the product (material or reactant) loss is minimised, as is the investment in treatment plants.

In establishing a new industrial plant, pollution control must be integrated from the beginning in the overall plant design. This is the best way to achieve maximum pollution prevention and optimum safety at minimum cost.

During site planning and plant lay-out, special attention must be given to rational installation, permitting proper management of pollutants and modern "husbandry". This applies, for instance, in the segregation of the different plant effluents as well as storm waters, so that optimum recycling, re-use or specific treatment can be carried out. The petrochemical study illustrates this approach. Both textile finishing and metalplating studies indicate that cramped sites and inadequate segregation of effluents are major factors in preventing the introduction of up-to-date pollution control techniques in factories.

Furthermore, pollution control equipment should be sited and operated as an integral part of the process unit. This ensures that pollution control "in plant" is managed with the same care and regularity as the production itself (and not as a "foreign body").

For industry, pollutant emission often means a net loss of valuable substances; either raw materials, reactants or finished products. Measures which increase efficiency of the manufacturing processes will prevent pollution as they save materials and improve the overall energy balance of the process. This approach, which undoubtedly is the most promising, is the basis of "clean" technologies. The choice of process used is of key importance.

The pilot studies have shown that a relatively simple step in pollution prevention is the careful choice of reactants, catalysts and materials in general; the scope is quite wide for textile finishing but relatively more limited for petrochemicals and metal plating. This approach also has the advantage of being applicable in existing plants with minimum adjustment. The market price of the various materials may limit the choice.

The potential for adjustment to manufacturing processes in existing plants will often be limited for physical and financial reasons. Nevertheless, a series of small changes may prove valuable (e.g. more efficient use of dyestuffs). In setting up a new plant, the scope is obviously much broader, and new technologies may be considered. Besides the water pollution aspect, other important factors such as possible transfer of pollution (into air or solid waste) or increased energy consumption will also need to be considered.

Recycling and re-use of materials and reagents are still limited but are becoming increasingly important in most activities under the combined pressure of both environment protection and scarcity of resources and materials. Recycling implies the reutilisation of a product for the same use, while reuse has a much broader sense and includes reutilisation for any purpose. These techniques can also be frequently introduced in existing plants and processes; for example, bath contents of many metal-plating processes can be recycled with relatively easy clean-up between cycles of use. In many other industries (e.g. sugar mills) recycling of process effluents has now become usual.

An approach applicable to most industries, which has proved very effective in pollution prevention, has been reduction in the volume of water used in the processes. This method, which generally requires numerous adjustments in the different phases of production, requires less wasteful processes, systematic recycling, reuse of waste waters and limitation of losses. Experience in several Member countries has suggested that water-saving techniques also result in a net pollution decrease. Moreover, with reduced volumes of effluent, treatment plants can operate more efficiently and be more compact, which means better pollutant removal and investment saving.

Finally, a conventional method of pollution reduction lies simply in good maintenance and husbandry throughout the plant, and good quality equipment; this guarantees safer operation and reduces chronic spills and accidental discharges generally caused by operational instability and negligence.

C. CONTROL OF FACTORY EFFLUENTS AFTER INDUSTRIAL PROCESSING

To improve the planning of the type and the degree of treatment needed, it is first necessary to consider the different alternatives for discharge of treated effluent:

i) The first involves direct discharge of purified effluent to natural waters without any further "barriers" to control possible pollution. This type of discharge is seldom advisable in industries covered by the pilot studies or similar ones, without supplementary equipment and security devices. These include: supplementary holding tanks of sufficient capacity; stand-by treatment installation for breakdowns; monitoring and safety devices to prevent accidental spills; physicochemical installations; and finally biological lagooning to ensure stabilization and polishing of purified effluent before its discharge to natural waters. The types of industry for which direct discharge of purified effluent presents minimum risk are the food and similar industries which have raw effluents with a high BOD load, high suspended solids and generally no toxic materials.

ii) The second and very common type is the discharge of industrial effluent to a (municipal or other) waste water treatment plant outside the factory. (To be examined below.)

iii) A third type is the discharge, through irrigation and spraying, to agricultural land (crops or pasture). Provided adequate preliminary geological and hydrological studies have been done and non-toxic effluent is spread at a rate compatible with vegetation and soil adsorption and purification rate, this method enables minimal environmental pollution and reuse of fertilizing elements. Control of eutrophication may also be obtained for effluents rich in nutrients. Nevertheless, precautions have to be taken to avoid any contamination of groundwater by specific pollutants, soil poisoning (by metals), water logging by excess spraying and discharge in winter when the soil is frozen or snow-bound. Where possible it is advisable to use the lagooning technique as a final stage before water is spread; this allows more flexible flow

regulation and effective removal of a good part
of the pollution load.

iv) A <u>fourth</u> type, the total or partial re-use of
the treated effluent for factory water, corresponds to a systematic recycling of waste waters
and is an excellent solution in terms of resource conservation and pollution protection.

On-site treatment of effluents by industry is a
basic barrier to the spread of pollutants, whether the
final effluent is discharged to natural waters or to
municipal sewers. Pre-treatment and full treatment can
be distinguished.

In general, the objective of pre-treatment is to
remove toxic or nuisance materials from effluents which
would interfere with further treatment either in the
municipal sewage treatment plant or at the industrial
central site treatment plant. Pre-treatment generally
is better carried out on each particular effluent stream,
thus permitting more specific treatment and more efficient
pollutant removal. In some particular cases, mutual neutralisation of effluent streams can be advantageous.

Full treatment in central sites varies from industry
to industry. Incoming pre-treated waste waters can be
balanced and mixed to equalise flow rates, concentrations
of pollutants, pH, temperature etc. The commonest type
of central site treatment involves biological processes
which offer various advantages in terms of cost, regularity and some guarantee of quality of the final effluent.
Nevertheless, as it cannot cope with all types of pollutants and may be poisoned by toxic waste, it may be
necessary to introduce supplementary physico-chemical
steps either before or after the biological treatment.
Finally, for direct discharge to natural waters, polishing either through physico-chemical processes (activated
carbon) or simple natural processes (lagooning) is
desirable.

In order to avoid acute shock effects to aquatic
ecology in the neighbouring discharge zone when a (purified) effluent is discharged to natural waters, it is
essential that pollution concentrations do not exceed
certain limits. Nevertheless, it should be realised
that the concentration is not the only factor to be considered; even more important for the river as a whole,
is the <u>total load</u> of each pollutant discharged every day.
In the past, a common method of treatment was to simply
dilute concentrated toxic effluents with relatively pure
waters (such as cooling waters) so as to reach the required standard. This procedure should be firmly discouraged as it is a means of complying with pollution
control regulations, which does not actually remove the

pollutants from the effluent and does not diminish the overall contamination of the water environment. It is, thus, essential that effluent discharge regulations and monitoring procedures be based on total discharge of pollutants and total flow of effluents per day in relation to corresponding industrial production.

D. CONTROL OF EFFLUENTS OUTSIDE THE FACTORY

Industrial effluents, raw or partially treated, may receive final treatment outside the factory, either in municipal plants or in specialised plants.

Under normal circumstances, municipal treatment plants (normally based on biological treatment) are designed to treat the gross pollution from domestic sources (BOD and suspended solids). An overall assessment shows that, in practice, removal of specific pollutants is quite problematic and coincidental; a relatively small proportion of pollutants are well removed, while a large number pass through the plant with low removal. The functioning of municipal plants is frequently disturbed or even interrupted by poisoning from toxic industrial waste. For these reasons, with the exception of those industries such as parts of the food industry whose effluents contain no toxic or harmful substances, it is necessary to pretreat the effluent at the industrial site to remove the toxic or other substances likely to disturb treatment in municipal plants, and the types of pollutants not satisfactorily eliminated by these municipal plants. Under these conditions, a well-operated municipal waste water treatment plant will offer an important supplementary safety factor vis-à-vis direct discharge to natural waters of industrial effluent treated only "on site".

In certain densely industrialised areas, there are waste water treatment plants specially designed for industrial effluents which receive the waste waters (pretreated or not) from neighbouring industries. These plants are often managed by industrial associations. Excellent equipment and highly qualified personnel are essential. The combination of different types of treatment (chemical, physical, biological) is necessary to eliminate the great variety of pollutants treated. The handling of sludges from such plants is, of course, a serious problem. In some countries in which polluting industries are scattered geographically and the transport costs for using a centralised plant becomes prohibitive, an alternative and already tried variation is to use mobile treatment plants mounted on trucks. These mobile plants then visit industries (such as metal-plating, tanneries etc.) to collect and treat their noxious wastes on site.

41

E. CONTROL OF MUNICIPAL EFFLUENTS

Municipal waste water treatment plants typically receives four types:

- domestic sewage which in general contains a limited load of specific pollutants from consumer products (detergent and cleaning products, pharmaceuticals etc.);
- effluents from commercial and workshop activities which may contain a relatively large variety of pollutants including toxic ones;
- effluents (pre-treated or non-toxic) from industry which may give rise to problems unless care is taken. If hazardous pollutants are still present they may be inadequately removed in the municipal plant and furthermore disrupt its operation by the poisoning of the biological treatment;
- urban run-off: storm run-off, especially after a long dry period, is rich in specific pollutants (metals, asbestos, lubricants, tars etc.). According to the types of sewer systems, storm run-off is either excluded from or partly discharged to treatment plants. Moreover, because of the types of pollutants generally encountered, their dilution and flow irregularity, present-day conventional treatment plants are not well equipped to remove the pollutant load from urban run-off.

The control of diffuse ("non-point") sources, and the monitoring and assessment of the load, distribution and pathways through the environment of the main pollutants involved are still at a preliminary stage. It is difficult to summarise the feasible control measures, but these do fall into two broad approaches:

i) the control at source of the causative activity by the most appropriate means (regulatory, managerial, technical). Examples include: limitation of airborne sources through stricter control of fuel combustion emissions; stricter control of agricultural sources (pesticides, fertilizers) through better management practices and stricter regulation of products and their use;

ii) the conversion, where possible, of "non-point" sources into a system manageable as a "point" source. For instance the control of urban run-off can be achieved through the proper management and treatment of storm waters.

42

Accidents and their pollution effects create a
serious problem; their range from purely unexpected
emergencies to neglect or even voluntary actions. In
this field a "multiple barrier" prevention approach is
likely to lead to essential progress. Much greater security for example in industrial operations, might often
be provided at relatively low cost by such safety devices
as holding tanks, by-passes, overflows etc. The adjunc-
tion of a few preventive "barriers" would frequently
transform a potential accident into an incident with far
less impact. In transportation, rules and codes of practice have been established in a number of international
conventions.

F. POTABLE WATER TREATMENT AS THE "ULTIMATE" BARRIER

Despite the different control "barriers" discussed
above, many pollutants still reach natural waters and
may result in a considerable load. These come from:
(i) residual pollutants in treated effluents after
treatment in municipal or industrial plants as a result
of the limited removal rate for many specific pollutants;
(ii) uncontrolled discharges, regular or irregular (industrial, domestic etc.); (iii) accidents; (iv) non-
point sources.

Because a residual load of non-degradable or difficult to degrade pollutants is now likely to build up in
rivers, regulatory authorities should apply stricter
pollution abatement on all controllable sources of these
pollutants. The previous idea allowing polluted discharges to saturate the so-called "assimilative capacity"
of the water bodies, is unacceptable because it precludes
the maintenance of water quality and is incompatible with
the other uses of natural waters including ecological
demands, recreation, and potable water supply.

The potable water treatment "barrier" has a special
place as it is the ultimate one before human consumption.
The treatment gives satisfactory results with raw waters
of reasonably good quality. When raw water quality decreases, the corresponding quality of drinking water may
become unacceptable as a large number of specific pollutants (soluble, colloidal, or in fine particles) pass
through the water treatment with a low rate of removal,
or combine with the reagents (organo-chlorines). Potable
water treatment should only be regarded as a last "security barrier" for raw water resources of good quality.
It should in no way be considered as a normal stage of
pollution control, in place of upstream waste water
treatment. This has not always been admitted in the past.

I.7. ECONOMIC FACTORS IN POLLUTION CONTROL

The contribution of economic analysis to pollution control policy mainly revolves around an analysis of the costs and the consequences of control costs to the particular company and industry involved and to the wider national or regional economy.

In developing a policy of pollution control for an industry, cost considerations are of key importance; a number of technological options exist for any particular environmental quality control but many of these can be disregarded because they are uneconomical. In economic terms the task of the environment decision-maker should be to establish a policy which would reduce pollution level to the point where increased pollution control costs are just offset by increased benefits (i.e. avoided damage cost to the environment); beyond this point the costs would exceed the benefits. This principle is only theoretical, since in practice this point is not actually known; moreover, the concept itself may be unacceptable as it might implicitly justify a "right" level of environmental damage, artificially based on an economic scheme.

In practice, the cost approach often adopted in the development of a pollution control policy for industry is that of cost effectiveness. Cost-effectiveness studies develop estimates of the minimum cost of achieving alternative levels of environmental quality and let the optimum level be decided through some political choice process. Industrial studies of this kind, investigating the trade-off between pollution control and the costs of control have been prepared by the OECD for several industries including textiles, metal-plating and petro-chemicals.

These studies can provide a framework for assessment of the cost impact of a chosen policy towards an industry and the wider economic aspects involving prices, employment, and international trade.

The method used in the studies was (a) to consider the raw residual loads associated with different unit processes of production; (b) to examine the available control technology; and (c) to examine the minimum costs of achieving given levels of production reduction. By accounting procedures the operating and capital costs of pollution control per unit of production could be simplified to a <u>single annual cost</u> figure, thus allowing a

straightforward cost-comparison between different levels
of pollutant removal.

The result of these studies expressed in cost/
effectiveness ratios, follows, in general, the expected
pattern according to which the cost of removal of an
extra unit of pollution increases as zero pollution is
approached. A number of factors which offset this effect
should however be noted. One is the relatively low cost
of good housekeeping practices and the diminution of
pollutants they make possible. Another is the adoption
of water conservation techniques: one set of calcula-
tions of a model metal-plating plant with advanced
counter-current rinsing techniques, working to capacity
on a restricted set of plating processes, does in fact
show a reduction in annualised costs at high levels of
pollution control when reduced end-of-pipe treatment and
water savings are taken into account. Water conservation and recycling require process modifications often
involving relatively major investment in plant to achieve
lower operating costs, but these investments may be offset by the savings in water charges and effluent
treatment.

In addition, increasing use is being made of newer
pollution control techniques, (e.g. reverse osmosis and
ion exchange) and there has been a tendency for many such
techniques to become more price-competitive as they pass
the experimental, model or pilot stage. Techniques for
economic recovery of raw materials appear promising, but
so far have not been widely used except in the case of
ions, e.g. the established methods for recovery of certain precious metals used in metal-plating and the
electronics industry. Centralised waste treatment plants,
which have been installed in a number of Member countries,
operate generally under market conditions and it has
been found difficult to recover from concentrated liquid
or solid waste economically.

Production capacity and plant utilisation factors
are important in assessing the cost-effectiveness of
pollution control investments. Important factors are:
(a) the overall production capacity of the plant, which
may vary with economic conditions; and (b) the degree to
which pollution equipment or plant modifications are
used.

Factors of scale are important in plant modification
and purchase of equipment; larger firms are at an advantage here. An earlier study on the iron and steel industry suggested that pollution control could account for
up to 20 per cent of total costs.

Although figures may vary widely according to type, size and age of the industry, available studies suggest that considerable reduction in pollution load can be achieved at relatively low cost (vis-à-vis production cost) and low level of investment (vis-à-vis total investment).

However, in the policy-makers' decisions on what pollution control policy to adopt and on the timetable for enforcement, the cost impact for each type of industry is an important consideration. Here such factors as profitability, availability of investment resources, possible loss of sales associated with any price rises resulting from pollution control, and international competitiveness are important.

Studies at a macro-economic level for several countries(*) have shown only slight effects of environmental programmes on prices, employment and gross national product (GNP). Although environmental regulations may contribute to the closure of some marginal plants with consequent regional employment effects, environmental policies also create jobs, e.g. in the anti-pollution equipment sector, and national analysis suggest the overall effects are neutral or positive. Although the benefit obtained from environmental policy - a cleaner environment (in the case of water, cleaner beaches, better fishing, better potable water) - are not reflected in GNP, the studies indicate that output as traditionally measured is not unfavourably affected by expenditure on the environment. Capital expenditure in the environment sector combined with associated technological innovation each make a useful contribution to growth in output.

(*) Summarised in "Macro-Economic Evaluation of Environmental Programmes", OECD, 1978.

I.8. CONCLUSIONS

A. THE PROBLEM OF SPECIFIC POLLUTANTS

Specific pollutants(*) present a long-term and growing problem which must be dealt with. They may, to a
large extent, pass through conventional potable and waste
water treatment plants. Such pollutants may be toxic,
persistent, bio-accumulated, and pose long-term health
hazards. Their precise environmental effects, however
are still often largely unknown, and action will have to
be taken on the basis of provisional assessments.

B. ADAPTATION OF WATER PROTECTION POLICIES IN RELATION TO
   THE EVOLUTION OF ENVIRONMENTAL CONTAMINATION

Action pursued in pollution abatement over the last
decade in Member countries has been mainly directed
towards the control of:

   i) traditional "gross" pollution (oxidisable
      matter/BOD, suspended solids);
  ii) pollution from major "point" sources (domestic
      and industrial sewers).

However it must be realised that, in general, the
problems of:

 iii) specific water pollutants (persistent toxic
      and bio-accumulated substances in particular);
      and

  iv) "non-point" (or diffuse) sources of pollution
      (from urban, agricultural and industrial
      origins)

_____

(*) The specific pollutants concept was used to
emphasize the difference between identifiable chemical
substances on the one hand and classical aggregate or
general parameters (such as BOD, COD and suspended solids)
on the other.

have simultaneously increased. They are likely to pose greater difficulties than (i) and (ii) and may call for the development of preventive and/or stricter policies.

C. WIDER USE OF MULTI-BARRIER STRATEGIES IN WATER POLLUTION CONTROL

A multi-barrier strategy is fundamental to a specific water pollutants control policy. It implies that a number of barriers (which may be of different orders i.e. technical, managerial, regulatory and economic) are placed simultaneously at different key phases which increases the total effectiveness of the system and takes account of the possible weaknesses of each barrier in preventing the passage of certain types of pollution. A multi-barrier approach will combine, for instance, stringent control at source (pre-production and pre-marketing control, strict product and use specifications) with cleaner technologies, recycling, early segregation of toxic waste, and finally treatment facilities. The concept is applicable at all levels of water management and pollution control (for example, in a river basin, factory or potable water treatment plant) and should always be kept in mind by regulatory authorities.

D. POLLUTION ABATEMENT THROUGH RECYCLING AND RE-USE OF WATER, MATERIALS AND REAGENTS

For industry, pollutant emission often means a net loss of valuable products - raw materials, reagents, finished products as well as energy (thermal pollution). Thus, measures which increase efficiency in the manufacturing processes will prevent pollution as they save water and materials, and are also likely to improve the overall energy balance of the system.

Recycling and re-use(*) of materials and reagents, until now relatively limited, are considerably increasing in most activities under the combined pressure of environment protection and resource scarcity (water, materials, and energy). These methods, which form the bases of clean technologies in new industrial processes, can often be introduced in existing plants and processes as well. The systematic reduction in the volumes of

---

*) Recycling generally implies the re-utilisation of a product for the same use, while re-use has a much broader sense and includes re-utilisation for any useful purpose.

48

water used in industrial processes (successfully undertaken in a number of industries) has also proven effective in decreasing the total pollutant load emitted in
industrial effluents. Reciprocally, pollution control
measures have often led to a reduction in the volumes of
water used in industrial processes.

E. THE NEED FOR LONGER-TERM PLANNING IN POTABLE WATER
   SUPPLY AND REORIENTATION OF POLICIES IN THIS PRIO-
                      RITY SECTOR

   Over the last two decades some deterioration in the
chemical quality of potable water has been noted in
various OECD countries, especially in regions of rapid
urban and industrial growth. Water demand has considerably increased in these areas during this period because
of growing population, increased consumption per capita,
and high demand from industrial and other activities
now supplied from the drinking water network for purposes
which do not require such good quality water. To meet
the ever-increasing demand, the existing (generally good
quality) traditional sources have frequently been supplemented by raw waters of low quality from "easy-to-get-at"
rivers near urban areas.

   The increased use of low quality raw waters for
potable water supply should, as far as possible, be discouraged as it generally leads to a less satisfactory
product from the standpoint of human health and taste.
A large number of potentially hazardous micro-pollutants
initially in solution, pass unremoved through the treatment plant, while relatively harmless pollutants may be
transformed into toxic chemicals (e.g. organo-chlorines)
by reaction with the treatment reagents. All possible
efforts should then be made to procure the necessary
supply of good quality raw waters locally, regionally,
or if necessary, from longer distances. When the total
quantity of good quality water is limited on a regional
basis, absolute priority should be given to potable water
supply. All existing water resources should be rationally reallocated according to a strict hierarchy of uses.
In view of water contamination by specific pollutants,
policies for potable water supply and distribution may
need careful re-examination. Plans and decisions should
be made primarily in the light of health and social requirements and not only from the restricted engineering
and economic viewpoints, or on a short-term basis as may
have previously been the case. More rational and longer-
term planning of potable water supply and distribution
would not only better fulfil the health and social duties,
but might therefore, in most instances, prove less costly
for society as a whole.

F. A COHERENT SYSTEM OF WATER QUALITY AND EFFLUENT STANDARDS AND REGULATIONS

Rational water quality management, including control of specific pollutants, may be based on a system of matching instruments (i.e. standards), which are interdependent and harmonized (Figure 1). Standards for potable water and bathing water are based on health criteria, and they determine, to a large extent, the ambient water quality standards (together with other ecological criteria). The ambient quality standards and the "best practicable means" for pollution control and manufacturing technologies will in turn affect the desired level of effluent standards in industrial branches. Effluent standards should be supplemented by appropriate economic, managerial or technical instruments. Despite good available experience, a number of Member countries still only use some of the elements of the system, and often without the necessary interconnections.

In order to attenuate the current divergence of views between those countries favouring strict and those favouring loose international standards, a compromise approach is proposed. The objective is to harmonize the common features and to facilitate the normal convergence of Member countries' environmental protection policies while guaranteeing fair industrial competition; this should contribute to better protection of international rivers, estuaries and seas. On the basis of the work of the Expert Group on Industrial Branch Strategies for Pollution Control, it is suggested that consideration be given to the preparation of minimum standards for respective industrial branches, which could be made nationally or regionally more stringent according to the use proposed for receiving waters(*). These should

---

(*) The receiving waters can be:
1(a) Continental surface waters in general (rivers, streams, channels etc.).
1(b) Rivers or lakes which are ecologically more sensitive.
2(a) Coastal and estuarial waters in general, including bathing areas (as these areas are generally stretched along all coastlines in Member countries).
2(b) Coastal and estuarine waters which are ecologically more sensitive.
- Intermediary media may be:
3(a) Sewer systems connected to waste water treatment plants specially designed for industrial effluent purification.
3(b) Sewer systems connected to a normal municipal treatment plant.
3(c) Sewer systems not connected to a treatment plant and discharging to the aquatic environment. This point is to be referred to Items 1 and 2 above.

Figure I.1

DEPENDENCE OF WATER QUALITY STANDARDS

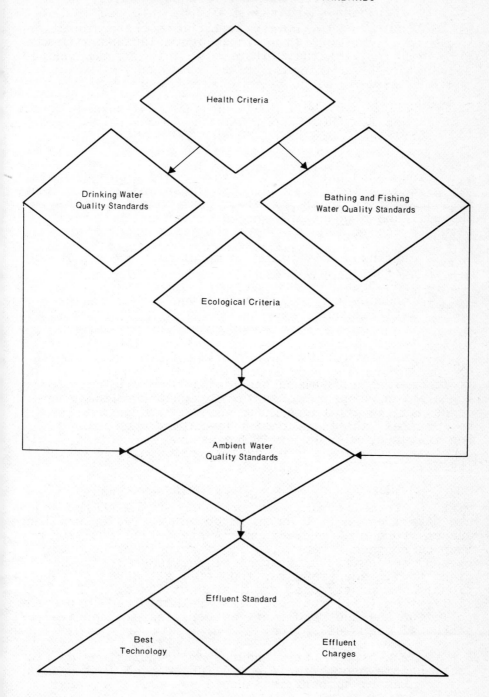

be based on effective industrial technologies which can
be reasonably applied in Member countries, and should be
regularly updated with the progress of these technologies.
For certain categories of hazardous pollutants, particularly stringent standards should be enforced on the
basis of their ecologically significant characteristics
(e.g. toxicity, persistence, bio-accumulation) with with a
view to preventing their dispersion into the environment.

G. THE NEED FOR EVOLUTIONARY AND PROGRESSIVE POLICIES IN ENVIRONMENTAL PROTECTION, AND FOR A MORE DYNAMIC AND SYSTEMATIC USE OF STANDARDS AND REGULATIONS

Experience shows that fixed and static regulations
may gradually lose their effectiveness and lead in turn
to a diminution in the level of environmental protection.
The well-known weakening and "erosion" of such regulations is due to several convergent factors such as:

- continuous evolution in manufacturing and other
  polluting activities;
- industrial growth and development;
- tendency for polluters to find ways round
  legislation;
- transfer and evasion of pollution via other
  forms or media;
- inflation and monetary erosion (charges, fines).

It is desirable to amend the static nature of certain
legislation which embodies fixed regulations and standards, and to include a dynamic procedure permitting
better flexibility and easier up-dating when necessary.
For instance, when a permit has been granted it is
necessary to allow for up-dating following a regular
review.

The lack of regulations and standards and the over-
rigid or formal conditions for their establishment and
use have frequently been an important obstacle in water
and environmental protection. A standstill position may
then be reached - hence (i) it is wrongly argued that
present scientific and technical knowledge is insufficient to fix any standard; thus (ii) standards or regulations, even provisional ones, are not fixed in this
field; then (iii) the polluters are not urged to prevent
pollution and the regulatory authorities are deprived of
a crucial means of action.

To make sure that standards and regulations do, in
fact, become more effective and fundamental instruments,
the regulatory authorities should see that they are:

52

- available for a larger range of sectors and
  parameters (even in a provisional form);
- established in the short-term on the basis of
  available knowledge;
- up-dated at appropriate intervals, and gradually
  strengthened as a result of: better knowledge,
  gradual adaptation of activities concerned and
  improved technologies and practices.

The establishment of standards should be accompanied
by a clear programme of implementation, fixing a schedule
for enforcing the different objectives.

H. MONITORING: A LIMITING FACTOR IN THE ENFORCEMENT OF ENVIRONMENTAL REGULATIONS

Monitoring of effluents, receiving waters, and
potable water is an integral part of water management
as it enables verification of whether imposed standards
and regulations are met. Nevertheless, the limitations
and the cost of monitoring methods are frequently a
handicap in implementing environmental legislation.
Chemical and physical analytical methods are now developed for a certain number of substances and parameters, but
for the vast majority of specific pollutants there is a
lack of knowledge. As many of these are still poorly
identified and associated with complex by-products (as
in effluents), "global" biological tests offer an
interesting solution, as they "integrate" all the ecological effects of pollutants without first requiring their
identification. Efforts are still required to fully
develop such biological tests which should be quick,
reliable, cheap and able to be routinely used within the
framework of international standardized methods.

I. STRICTER CONTROL OF POLLUTING PRODUCTS

The joint use of product-orientated and discharge-
orientated legislation is becoming a key feature in many
present-day national policies.

A frequent problem in effluent treatment is lack
of knowledge of the composition of products and reagents
used in industry, making their treatment difficult; some
compounds are particularly stable and toxic, while others
have hazardous by-products.

National and international regulations should seek
to impose:

- control over the toxicity and biodegradability
  of these products, taking into account the risk
  of formation of dangerous by-products or breakdown
  substances;
- a requirement to provide detailed qualitative and
  quantitative descriptions of all the chemical components of the product (at least to an approved
  central agency, to be used in confidence);
- a notice giving advice about recycling, treatment
  or detoxification of components.

Part II

ASSESSMENT, CATEGORISATION AND HAZARD RATING OF SPECIFIC WATER POLLUTANTS

II.1 <u>INTRODUCTION</u>

The OECD experts examined different systems for col-
lecting and structuring information on water polluting
properties of contaminants in order to develop one that
would best meet the needs of water management authorities
in taking decisions relating to pollution control.

The proposed system is a checklist for those primary
responsible for assessing the potential hazard of environmental contaminants and those who have to verify that
assessments have been carried out appropiately.

This section of the report has three main functions:
it discusses the scope of the required information and
outlines a method for its structuring; it lists pollutants by chemical class, source and intended function or
use; and it proposes a system for rating the hazard posed
by specific water pollutants.

II.2 SCOPE OF INFORMATION REQUIRED

The evaluation of the potential hazard of specific water pollutants requires that different kinds of qualitative and quantitative information be made available for analysis. In developing large systematic schemes for minimising the effects of new or established contaminants on environmental quality, it is recommended that the information be structured in a way that facilitates comparison, achieves consistent evaluation, and furthermore, prevents oversight of factors, the importance and relevance of which may not always be initially evident.

A. GENERAL PRINCIPLES

The information required for evaluating the potential impact of pollutants should cover the following main items:

- chemical class and structure;
- basic physical and chemical properties including biodegradability;
- the degradation products resulting from biodegradation of the original pollutants and their antagonism/synergism;
- anticipated total production;
- sources and channels of distribution, purpose(s) and pattern(s) of use, including the quality and flow of the receiving water and its downstream uses;
- practical situations of discharges to water bodies;
- quantities involved in identified discharge situations;
- toxic or other adverse effects of the compound on water quality and ecology (persistence and bioaccumulation);
- availability of technical control measures.

It is advantageous to carry out the proposed evaluation using a stepwise procedure. This will mean that the need for further elaborate and time-consuming studies is recognised at the most appropriate stage. The order of the above items is intended to correspond with this stepwise approach. The first six represent information which is frequently readily available or can be obtained from

57

deskwork predictions. From analysis of this information the probability of the considered pollutants appearing in any kind of aquatic system in significant quantities will be known with reasonable accuracy. Thus, it will be possible to express the questions raised under the last three items in the form of a realistic programme i.e. studies and tests of possible effects on aquatic ecosystems, and conditions for controlling discharges.

Some Member countries are in the process of preparing, or have already enacted, legislation relating to the production and use of chemical products which may be hazardous to public health or to the environment. These laws, based on similar considerations to those for pharmaceuticals and pesticides, will often require that chemicals in general are evaluated in appropriate detail with respect to possible effects on different elements of the environment, including man.

The method of structuring information, which is outlined in the paragraphs which follow, is restricted to impacts on aquatic systems. However, a few modifications and additions will suffice to make it cover other environmental sectors as well as aspects of the interference of contaminents with public health. In the final section, contaminants which may be present in effluents locally or generally, and which may be rated as potential or established specific water pollutants are described by chemical class, source and intended function or use. These classifications are illustrative, not exhaustive.

B. EVALUATION GUIDELINES

a) <u>General properties</u>

i) <u>Chemical class and structure</u>

The chemical class and structure may provide, by analogy, some indication of the possible environmental activity of a compound. The chemical structure will sometimes have to be reported with some degree of approximation, for instance, for products and by-products of chemical and petrochemical industries, such as surface active agents, solvents etc. It should be noted here that impurities, for instance, unreacted starting materials or products of side reactions, can be of great environmental significance.

ii) Data on physico-chemical properties

Data on the physico-chemical properties of a compound may be used tentatively to predict its behaviour in an aqueous medium, e.g. its tendency to be precipitated, to diffuse into lipophilic systems, to volatilise or to become adsorbed on solids. Thus such data provide a key to its transport and final distribution between the different constituents of an aquatic ecosystem. The chemical reactivity is indicative of the likelihood of harmful interactions with exposed ecosystems. For generally unstable compounds, the products of decomposition will have to be evaluated as well as the original compound.

Physico-chemical properties like solubility, tendency to become adsorbed or to precipitate will have to be taken into account in the design of tests for the evaluation of properties such as toxicity and biodegradability. This is also valid when the results of the tests are to be interpreted for application to practical situations.

Physico-chemical data are normally readily available with the following being particularly important:

- solubility in water and in fatty matter;
- volatility;
- chemical reactivity;
- tendency to become adsorbed on the different types of solids found in aquatic sytems.

iii) Analytical

The specificity, sensitivity and accuracy of available methods for the quantitative estimation of a considered compound in relevant aquatic systems should be reviewed. Indirect methods of analysis may have to be relied on. The availability and sensitivity of automated monitoring techniques should be considered.

iv) Biodegradability

The biodegradability of a compound under relevant conditions is usually of focal interest in evaluation of environmental safety. Although standard procedures have been developed and officially recommended for certain groups of chemicals, for instance anionic and non-ionic surfactants, several factors prevent the adoption of single, universally applicable method for biodegradability estimates. The choice of method will frequently have to depend on the characteristics of the individual substance and situation. However, a long-term objective of Member countries should be to achieve agreed methods of analysis giving compatible and comparable results of the necessary accuracy.

b) Quantities in use and use patterns

i) General remarks

The probability that a compound will cause damage
to the environment does not depend exclusively on any
proven toxic or otherwise hazardous characteristic. The
conditions of its manufacture, storage, transport and use
are of similar importance and should accordingly be taken
into account when decisions relating to its hazardous
properties and probable environmental impacts are taken.
Furthermore, attention should be paid to by-products
from manufacturing processes.

ii) Sources

Both the primary producer (or point of generation)
and any secondary industrial or other user should be
identified. It should also be noted whether the use or
generation is widespread or restricted to a few locali-
ties only.

iii) Production and consumption data

Qualitative and quantitative predictions regarding
the final distribution of a contaminant in the environment require approximate data of the tonnages produced
for different types of consumption. From the management point of view, it may be sufficient to state the
tonnage produced and consumed as, for instance, "small"
or "significant" in comparison with the consumed tonnages of similar materials. When a compound is being
used for several purposes the relative importance of
these from the pollution point of view is noted when
possible.

When the compound in question is a by-product or
is of a generally uncertain origin, even a rough estimate
of the quantities involved may require some practical
case studies to be carried out.

iv) Use patterns

An analysis of the conditions under which a compound
is intended to be used, i.e. the purpose(s) and pat-
tern(s) of use, will improve the accuracy of predictions
regarding the discharge of it to waters in quantities
which might be of environmental significance. By such
an analysis, the uses which rarely result in an uncontrollable dispersion of the compounds in the environment
would be distinguished from those which result in discharges which are difficult to control and which,

accordingly, will have to be examined for their potentially important harmful effects on the environment.

v) Characteristics of established discharges

From the analysis of use patterns, it is possible
to derive information about how established discharges
occur, i.e. occasionally, regularly or continuously. It
will also be determined whether the discharges are restricted to a few or to a great many point sources which
in principle can be controlled, or whether the sources
are widespread and non-controllable so that they must be
characterised as diffuse.

In this context the primary and, when relevant,
secondary aquatic recipient of the compound is stated
(an installation for effluent treatment being considered
as a primary recipient).

vi) Loadings

Loadings of aquatic recipient systems can be presented as:

- quantities discharged per unit of time;
- quantities discharged per unit of production or
  the equivalent;
- expected concentrations in receiving waters,
  usually under conditions of minimum flow;
- or, some appropriate combination of these loading
  parameters.

The management problem influences the chosen
method - the first method of presentation is the most
appropriate when the overall situation to be described
is in a river basin; the second when similar industrial
sources are to be compared with respect to technical
standards; and the third when the risk of specified environmental effects, e.g. damage to fisheries or effects
on water supplies, are to be assessed.

c) Environmental effects

i) General remarks

The direct or indirect effect(s) of a compound on
aquatic systems may be:

- impact on the receiving waters, their ecology and
  their responses, including acclimatisation, to
  the pollutant;

- reduction of the quality of the receiving water in relation to its use for human consumption, recreation, fish culture, agricultural and industrial supply;
- effect on processes for preparation of potable water and effects on distribution system;
- disturbance of biological, chemical or physical processes for treatment of effluents, including associated operations and damage to sewer systems; or
- a combination of the above effects.

Important aspects of the information on these effects are summarised in the hazard rating proposed in II.5. Laboratory experiments may sometimes be inadequate for fully predicting real environmental and health impacts because antagonistic and synergistic effects in particular, which frequently occur in the complex aquatic medium, are rarely predictable.

In the near future rapid and direct screening of pollutants for other hazards may well be brought into normal practice. Tests are being intensively developed for cytotoxicity, mutagenicity, carcinogenicity and teratogenicity. The perfection and evaluation of these indicative tests still requires much careful effort and confirmation of their validity through epidemiological studies.

ii) <u>Toxicity to man</u>

Within the context of water management, the potential long-term effects of contaminants on man require attention in cases where sources of potable water are affected, in particular by effluents of industrial origin. Direct uptake and uptake <u>via</u> foodchains must be taken into account. Also, consideration has to be given to the initial pollutants themselves and to their metabolites, and to any derivative that may be formed from them when reagents such as chlorine or ozone are applied. Finally, consideration must be given to compounds added during purification processes (as the reagents themselves or impurities in them).

It should also be noted that there are several tests, which assay groups of similar pollutants and which can be used rapidly to get a preliminary indication of hazard/toxicity before a series of tests for particular substances is undertaken. Typical examples are "total organic carbon" before tests for individual pesticides, and "ether extractable material" before tests for individual polar compounds. Great caution should be exercised, however, in doing these "blanket" tests where several groups of compounds are present.

iii) Toxicity to micro-organisms

Tests of acute toxicity to micro-organisms are particularly relevant when a contaminant is likely to be discharged at a reasonable concentration to plants for biological treatment of effluents. It should be noted that toxic effects of a compound towards bacteria, protozoans etc., as demonstrated under test conditions, may present different aspects in practice due to factors such as degradation, adsorption, recombination or complex forming.

iv) Toxicity to fish

The LC_{50} for a contaminant (concentration being lethal to 50 per cent of an exposed population of test fish within a given time) is a basic parameter in the present context. For estimation of LC_{50} values, various procedures using different test species and experimental conditions exist in the literature: the reported results for the same chemical compound may consequently vary within certain limits. No procedures have gained wide international acceptance so far. However, among others a special sub-committee on biological tests for water is working actively on internationally acceptable standard tests.

Because standardized tests for toxicity of pollutants to fish and other aquatic organisms are not yet fully developed, and because of the variability of reported data, it is useful to obtain another type of assessment based on ranges of concentration and properties of the pollutant as proposed in II.5.

v) Degree of physico-chemical or biochemical persistence

This point covers information relating to physico-chemical or biochemical processes by which compounds are deactivated in or eliminated from aquatic systems, either spontaneously or through planned measures. It also includes converse information relating to their stability (persistence) in an aquatic environment. Since the relevant data may originate from laboratory tests, their applicability to practical conditions may be difficult to evaluate precisely. Persistence and stability may therefore have to be stated in approximate terms, for instance as "low", "medium" or "high".

vi) Bioaccumulation

In ratings of specific water pollutants, the tendency of a compound or of one of its derivatives or

63

metabolites, to become concentrated in persistent form in living organisms ("bioaccumulation") is of key importance. If the risk of bioaccumulation cannot be established experimentally, it may be predicted by considering its physico-chemical properties and stability. It may be noted that a persistent compound need not have bioaccumulating properties, while a bioaccumulating compound is generally highly persistent.

Although quantitative methods are not fully developed, there are several test procedures being perfected at present. Until these methods are finalised quantitative data from them should be interpreted with caution, especially since data from different sources are not necessarily comparable.

Bioaccumulation is only one particular type of accumulation of pollutants in nature, and other instances of accumulation such as occurs in layers of lake water, or in soil strata may also be environmentally significant. Data on such phenomena can give important indications of the behaviour of pollutants.

d) <u>Technical control measures</u>

The final part of the overall evaluation procedure consists of an examination of the existing or potential technical barriers against undesirable dispersion of a pollutant in the environment. The conditions for removal at source (by generally practised processes for effluent treatment or by selective methods), at plants for collective treatment of industrial and domestic effluents, and at potable water works should be evaluated.

Information regarding physical and chemical properties and biodegradability, gives sometimes a good indication of the removability of a pollutant by one or more of the unit processes and operations which are in common use in effluent and potable water treatment. When available information or simple analogy do not permit predictions of removability, detailed technical considerations and experiments might be required.

II.3 <u>DESCRIPTION OF POTENTIAL WATER POLLUTANTS</u>

The following Tables index pollutants by:

A. Chemical Class
B. Source
C. Function

These Tables have been developed to illustrate the assessment procedure outlined, and are related to the other parts of the programme on water pollutants control. Pollutants are listed in the form likely to be present in river water or effluents.

The lists are only indicative, and not comprehensive. They exclude many substances, especially those found at low concentrations or in small amounts. However, omission from the list does <u>not</u> imply that the compound is not hazardous.

Table II.1

POTENTIAL WATER POLLUTANTS

Chemical Class

Class of Pollutant	Example	Sources	Functions
ALCOHOLS	Methanol	10, 12	41, 72
ALDEHYDES	Formaldehyde	8, 10, 12, 16, 17, 24	47, 65, 74
	Glyoxal	33	
ESTERS	Dibutyl phthalate	4	59
HYDROCARBONS	n-Alkanes	2, 7, 17, 26, 32, 33	36, 41, 44
	Styrene	29	47
	Naphthalene	3, 15, 21	13
	3, 4 Benzfluoranthene	21, 32	41
	3, 4 Benzpyrene	21, 32	41
KETONES	Acetone	9	71
	Methyl iso-butyl ketone	9	71
ORGANO-HALOGEN	Ethyl chloride	21	
	Vinyl chloride	13	47
	1, 2 Dichlorethylene	13	13
	Bis chloroisopropyl ether		
	Chloro-paraffins	7	22
	DDT	1, 5	56
	Dieldrin	33	49
	Eulan	33	49
	Pentachlorophenol	12, 27, 33, 34	37, 64
	Polychlorinated biphenyl	20	37, 56
	Atrazine	1	39
	2, 3, 6 Trichlorobenzoic acid	1, 5	39
	2, 4 Dichlorophenoxyacetic acid	1, 5	39
	Trichlorobenzene	33	
	Hexachlorobenzene	14	13
ORGANO-PHOSPHORUS	Parathion	1	56
ORGANO-METAL	Tri-butyl tin oxide	1	56
ORGANO-NITROGEN	Acrylamide	36	47
	Acrylonitrile	24, 29	47
	Carbamates	1	37
	Dimethylamine	16	74
	EDTA	4, 17, 27	70
	NTA	4, 27	70
	Urea	1, 2, 12, 14, 24, 32	23, 32, 60
	Aniline	3, 17	13, 43
	Nitrobenzene	3, 5	43
	Nitrophenols	8	31
	Nitrocresols	1	56
	Pyridine	3, 5	13, 43, 71
	Morpholine	5	43
	Para-amino stilbene type dyes	4, 33	27, 52
	Cetyl trimethylammonium bromide	11, 33	15, 25
ORGANO-SULPHUR	Carbon disulphide	35	71
	Sulphurised hydrocarbons	7	22
	Zanthates	35, 18	34, 71
	Thiophene	21	13, 41
	Dodecyl mercaptan		51
	Alkyl benzene sulphonates	33, 4	5
	Methylthiobenzothiazole	24, 29	8
PHENOLS	Phenol	3, 7, 10, 15, 21, 33, 34	13, 25, 27, 29, 60, 64
	Lignin	27	13
	Orthophenyl phenol	33	27
	Cresols	24	60

66

Table II.1 (cont.)

Class of Pollutant	Example	Sources	Function
SALTS	Formate	13, 16, 33	27
	Acetate	17, 20, 24, 28, 30, 33	13, 27, 60, 69
	Propionate	28	13, 27
	Stearate	4, 17, 30, 33	73
	Naphthenate	21	13, 64
	Oleate	4, 17, 30	73
	Palmitate	30	73
	Oxalate	17, 33	27
	Tartrate	17	20
	Peracetate	33	54
	Heptonate	17	20
	Gluconate	17	20
	Benzoate	33	
	Salicylate	33	
	Phthalate	16	
STEROIDS	Cholesterol	36	45
	Ethinyloestradio	4	21
	Mestranol	4	21
	Norethisterone	4	21
	Norgestrel	4	21
METALS & METALLOIDS	Aluminium	15, 17, 18, 27, 33	13, 18, 41, 48, 74
	Antimony		
	Arsenic	1, 17, 18, 20	11, 41, 55, 56
	Barium	7, 17, 20	38, 55
	Beryllium		57
	Cadmium	17-20, 31, 32	41, 46, 55
	Calcium	1, 12, 14-18, 27, 36	3
	Chromium	6, 15, 16, 17, 20, 26, 33	46, 54, 55, 58
	Cobalt	17, 20	55
	Copper	1, 6, 7, 17, 18, 19, 20, 26, 33	27, 37, 41, 46
	Iron	6, 15, 17, 18, 20, 23, 26, 28	41, 46, 55
	Lead	8, 15, 17, 18, 19, 20, 21, 31, 32	7, 28, 41, 46, 53
	Manganese	15, 17, 38	41, 46
	Magnesium	14, 15, 18, 24, 27	18, 41
	Mercury	1, 8, 14, 19, 27	28, 37, 41, 44, 46
	Molybdenum		
	Nickel	17, 21, 31	28, 41, 46
	Potassium	1, 17	13, 69
	Silver	22	
	Sodium	1, 2, 4, 6, 7, 11, 14, 16, 17, 18, 21, 22, 25, 27, 30, 32, 33, 35, 36	3, 41, 69
	Strontium	7	38
	Tin	1, 17, 18	46, 53, 56
	Titanium		46, 55
	Uranium		
	Vanadium	21	4, 41
	Zinc	1, 4, 7, 15, 17-21, 29, 32, 35	4, 14, 35, 41, 46, 53, 55
CATIONS	Ammonium	1, 2, 3, 7, 8, 10, 14, 15, 17, 24, 27-30, 36	13, 14, 35, 32, 41, 69
ANIONS	Borate	4, 16, 17, 22	10, 69, 74
	Chloride	1-4, 6, 7, 11, 12, 14-18, 20, 25-28, 30, 32, 33, 36	3, 10, 13, 23, 30, 41, 69
	Chlorate	1	39
	Cyanide	3, 7, 15, 17, 20	13, 20, 38, 55
	Ferricyanide	20	13, 55
	Ferrocyanide	20	13, 55

Table II.1 (cont.)

Class of Pollutant	Example	Sources	Function
ANIONS (cont.)	Fluoborate	17	20
	Fluoride	6, 14, 15, 17, 19, 29	14, 30, 41
	Fluosilicate	17	20
	Nitrate	8, 14, 17, 22, 33, 36	2, 13, 31, 32
	Nitrite	2, 7, 17, 36	13, 67, 22
	Phosphate	1, 4, 11, 12, 14, 15, 17, 36	2, 13, 17, 32, 67, 68, 70
	Polyphosphate	4, 11, 17, 27, 33	17, 70
	Polysulphide	15, 17	13
	Silicate	4, 11, 17, 29, 33	3, 17, 35
	Sulphamate	17	
	Sulphate	1, 4, 8, 12, 14, 15-21, 24, 27, 28, 31, 33, 35, 36	2, 13, 19, 69, 41
	Sulphide	3, 15-17, 20, 21	19, 53, 55, 75
	Sulphite	12, 22, 27, 33	19, 25, 65
	Thiocyanate	3, 15	13
	Thiosulphate	3, 15, 16, 22	13, 20, 65
SOLIDS	Asbestos		12

Table II.2

POTENTIAL WATER POLLUTANTS BY

Source

Source	Inorganic		Organic
	Cations	Anions	
1. AGRICULTURE			
1.1 Fertilizer use	Ammonium Calcium Potassium Sodium	Chloride Phosphate Sulphate Nitrate	Urea

1.2 Herbicide use

A wide vareity of herbicides may be used. Washing of plant and equipment and careless disposal of containers may result in pollution of water. For list of substances see Approved Products for Farmers and Growers, 1973, Ministry of Agriculture, Fisheries and Good, London.

Examples are:			2, 4 D 2, 4, 5 T Atrazine etc.

1.3 Pesticides and fungicides

A wide variety of pesticides may be used. Those of a persistent nature (such as the chlorinated hydrocarbons) may be leached from soil into surface waters.

The pesticides used may include the following classes:

Source	Cations	Anions	Organic
	Arsenic Copper		Carbamates Nitrocresols etc. Organo-chlorine Organo-phosphorus Organo-mercury Organo-tin Thiocarbamates
1.4 Animal excreta	Ammonium Copper Potassium Zinc	Phosphate	
2. AIRPORTS			
2.1 De-icing	Ammonium Sodium	Chloride	Alcohol Urea
2.2 Aircraft cleaning		Nitrite	Alcohols Oil Surfactants
3. CARBONISATION			
3.1 Coal gas, coking etc.	Ammonium	Chloride Sulphate Sulphide Thiosulphate	Aniline Cyanide Naphthalene Phenols Pyridine Thiocyanate
4. CONSUMER PRODUCTS			
4.1 Detergents	Sodium	Borate Chloride Phosphate Polyphosphate Silicate Sulphate	Anionic surfactant Cationic " EDTA Non-ionic surfactant NTA Oleate Stearate
4.2 Oral contraceptives			Ethinyl oestradiol Norethisterone Norgestrel and others Mestranol
4.3 Cosmetics	Zinc		Phthalate esters

Table II.2 (cont.)

Source	Inorganic		Organic
	Cations	Anions	
4.3 (cont.)			
A wide variety of inorganic and organic chemicals is used in toiletries, cosmetics, medicines, cleaners, paints, etc. in the home and may become water pollutants			
5. DYESTUFFS, INTERMEDIARIES & FINE ORGANIC CHEMICALS			
Potential pollutants include a very wide range of inorganic and organic acids, bases and salts, reducing and oxidising agents and unknown side reaction products.			
Examples are:		Chloride Nitrate Nitrite Sulphate	Aniline Morpholine Nitrobenzene Pyridine
6. ELECTRONICS			
6.1 Printed circuit manufacture	Copper Iron Sodium	Chloride Chromate Cyanide	
6.2 Transistor manufacture		Fluoride	
7. ENGINEERING			
7.1 Lubrication and cooling of cutting tools		Nitrite	Bactericides Chlorinated hydro-carbons Oil Sulphurised hydro-carbons Surfactants Polyglycols Phenols Water miscible solvents
7.2 Hardening	Barium Strontium Sodium		Cyanide
7.3 Soldering	Ammonia Copper Zinc	Chloride	
8. EXPLOSIVES			
8.1 Propellents Blasting etc.	Ammonium	Nitrate Sulphate	Formaldehyde Hexamine Nitro- and nitrated organic compounds (TNT, nitroglycerine picric acid)
8.2 Detonators	Lead Mercury	Azide	
9. FERMENTATION			
9.1 Antibiotics			Acetone Antibiotics Methyl isobutyl ketone
10. FIBREGLASS			
10.1 Insulation	Ammonium		Formaldehyde Methanol Phenols Resin

Table II.2 (cont.)

| Source | Inorganic | | Organic |
	Cations	Anions	
11. FOOD PROCESSING			
11.1 Cleansing	Sodium	Chloride Hypochlorite Phosphate Polyphosphate Silicate	Cationic surfactants Iodophors
11.2 Lye peeling	Sodium		
12. GLUES AND ADHESIVES			
12.1 Urea/formaldehyde			Formaldehyde Methanol Urea
12.2 Glue and gelatine	Calcium	Chloride Phosphate Sulphate Sulphite	Gelatine Ossein Pentachlorophenol
13. HEAVY ORGANIC CHEMICALS & PETROCHEMICALS			

Potential pollutants include a wide range of inorganic and organic acids, bases and salts, and solvents and various reaction products.

| Source | Inorganic | | Organic |
	Cations	Anions	
An example is:			
13.1 Vinyl chloride production			Chlorinated poly- merization products of low molecular weight 1, 2 dichloro ethylene Formate Vinyl chloride
14. INORGANIC CHEMICALS INCLUDING FERTILIZERS			
14.1 Sulphuric acid production	Mercury(⌀)		
14.2 Chlorine production	Mercury(⌀) Sodium	Chloride Hypochlorite	Hexachlorobenzene(*)
14.3 Sodium carbonate production	Calcium Magnesium Sodium	Chloride Sulphate	
14.4 Fertilizers manufacture	Ammonium Calcium	Fluoride Nitrate Phosphate Sulphate	Urea
14.5 Phosphorus production	Phosphorus (elemental) (⌀) from pyritic ores (⌀) from cathode		(*) from anode
15. IRON AND STEEL (Including STAINLESS STEEL)			
15.1 Pickling	Iron Manganese	Chloride Fluoride Phosphate Sulphate	Inhibitors Surfactants
15.2 Gas washing (Blast furnace gas: coke-oven gas)	Ammonia Lead Magnesium Zinc	Chromate Cyanide Fluoride Sulphide Thiocyanate	Naphthalenes Phenol

Table II.2 (cont.)

Source	Inorganic		Organic
	Cations	Anions	
15.2 (cont.)			
15.3 Slag (leaching by rain)	Calcium	Sulphate Fluoride Polysulphide Sulphur Thiosulphate	Phenols
16. LEATHER			
16.1 Beamhouse	Calcium Sodium	Chloride Sulphide Thiosulphate	
16.2 Tanning and finishing	Aluminium Chromium	Borate Sulphate	Dimethylamine Dyestuffs Formaldehyde Formate Phthalate Sulphonated oils Surfactants Syntans
17. METAL FINISHING			
17.1 Degreasing	Copper Zinc	Cyanides Phosphates Silicates	Amines EDTA Oil Surfactant Trichlorethane Trichlorethylene
17.2 Pickling	Chromium Copper Iron Manganese Sodium Tin Zinc	Chloride Chromate Cyanide Fluoride Nitrate Phosphates Silicates Sulphate	EDTA Gluconate Heptonate Surfactant
17.3 Wet polishing	Ammonium	Borate Chromate Cyanide Phosphate Silicate	Corrosion inhibitors Sequestering agents Soap Surfactant Tartrate
17.4 Chemical and electro-chemical deposition (electroplating)	Cadmium Copper Lead Nickel Tin Zinc Sodium	Chloride Chromium Cyanide Borate Fluoride Phosphates Sulphate Sulphamate Fluoborate Fluosilicate	EDTA Formaldehyde Hydrazine Various additives
17.5 Brightening, polishing, passivating etc.		Chromate Nitrate Phosphate Sulphate	Aniline Glycerine
17.6 Bright dipping	Cadmium Copper Zinc	Chromate Nitrate Sulphate	
17.7 Chemical colouring e.g. bronzing, blueing	Arsenic Ammonium Calcium Copper Lead Potassium Sodium Nickel	Chromate Phosphate Polysul- phide Sulphide	

Table II.2 (cont.)

| Source | Inorganic | | Organic |
	Cations	Anions	
17.8 Anadozing & dyeing	Aluminium Ammonium Cobalt Iron Nickel Manganese	Chromate Sulphate	Acetate Dye Stuffs Oxalate
17.9 Rustproofing (including phosphating)		Chromate Phosphate	Hydrocarbons Ketone
17.10 Painting	Pigments		Oils Solvents
18. MINING			
18.1 Coal washing		Sulphate	
18.2 Coal storage	Iron	Sulphate	
18.3 Mine waters	Aluminium Calcium Iron Magnesium Manganese Sodium	Chloride Sulphate	
18.4 Tin	Arsenic Cadmium Copper Iron Lead Tin Zinc		
18.5 Lead/zinc	Aluminium Copper Lead Potassium Zinc	Silicate Sulphate Sulphide	Methyl isobutyl carbinol Oleate Surfactants Xanthates
18.6 Salt/potash		Chloride	
19. NON-FERROUS METALS			
19.1 Smelting	Cadmium Copper Lead Mercury Zinc	Fluoride Sulphate	
20. PAINTS AND PIGMENTS			
20.1 Titanium dioxide	Iron	Sulphate	Polychlorinated biphenyls
20.2 Various inorganic pigments	Arsenic Barium Cadmium Cobalt Copper Iron Lead Zinc	Chloride Chromium Cyanide Ferrocyanide Sulphate Sulphide	Acetate
21. PETROLEUM			
21.1 Refining	Nickel Sodium Vanadium	Sulphate Sulphide	Hydrocarbons Mercaptans Naphthenates Phenols Thiophene
21.2 Distribution	Lead Zinc		Ethyl chloride Hydrocarbons

Table II.2 (cont.)

Source	Inorganic		Organic
	Cations	Anions	
22. PHARMACEUTICAL			
Potential pollutants include a very wide range of inorganic and organic acids, bases and salts, reducing and oxidizing agents and unknown reaction products.			
Examples are:	Ammonium Calcium Magnesium Sodium	Nitrate Phosphate Sulphate	Acetone Acetate Mesityl oxide
23. PHOTOGRAPHIC			
23.1 Film Manufacture	Silver	Nitrate	Trace organics
23.2 Processing	Iron Sodium	Borate Sulphite Thiosulphate	Organic reducing agents and other chemicals
24. PLASTICS AND SYNTHETICS			
Potential pollutants include a wide range of inorganic and organic chemicals			
Examples are:			
24.1 Phenol/formaldehyde	Ammonium		Cresols Formaldehyde
24.2 Urea and melamine			Formaldehyde Melamine Urea
24.3 Cellulose acetate	Magnesium	Sulphate	Acetic acid
24.4 Acrylics			Acrylic monomer Surfactants
25. POWER GENERATION			
25.1 Water treatment	Sodium	Chloride	Inhibitors
25.2 Hydraulic transport of PFA	Trace metals		
26. PRINTING			
26.1 Offset litho Photogravure etc.	Copper Iron Pigments	Chloride Chromium	Oil
27. PULP AND PAPER			
27.1 Cooking	Ammonium Calcium Magnesium Sodium	Sulphate Sulphide Sulphite	Organic sulphur Terpenes Waste lignins
27.2 Bleaching	Sodium	Chloride Polyphosphate	EDTA DTPA
27.3 Paper manufacture	Aluminium	Chloride Hypo- chlorite Sulphate	Bactericides (Organo-mercury) Dyestuffs Polymeric retention aids (acrylamide)
28. REFUSE DISPOSAL			
28.1 Landfill	Ammonium Iron	Chloride Sulphate Sulphide	Acetate Propionate
If liquid, semi-liquid and solid toxic wastes are disposed of in landfill leachate may contain a wide range of potential pollutants.			

Table II.2 (cont.)

Source	Inorganic		Organic
	Cations	Anions	
29. RUBBER			
29.1 Latex (synthetic)	Ammonium	Silicates	Acrylonitrile Anti-oxidants e.g. Methyl thiobenzo- thiazol Bactericides Styrene
29.2 Foam	Zinc	Fluoride	
30. SOAPS AND DETERGENTS			
30.1 Soaps	Sodium	Chloride	Acetate Palmitate Glycerol Oleate Stearate
30.2 Surfactants	Sodium	Sulphate	Surfactants
31. STORAGE BATTERIES			
31.1 Nickel-cadmium	Cadmium Nickel	Sulphate	
31.2 Lead	Lead	Sulphate	
32. STORM WATER			
32.1 Road drainage	Cadmium Lead Sodium Trace metals Zinc	Chloride	Oil Polynuclear aromatics Urea
33. TEXTILES (Wool)			
33.1 Raw wool scouring (Woolcombing)	Sodium		Lanoline Nonyl phenol ethoxylate Stearate
33.2 Yarn scouring			Nonyl phenol ethoxylate Oil
33.3 Wool dyeing	Sodium	Sulphate	Acetate Formate Dyestuffs Softners (cationic) Surfactants
33.4 Mothproofing			Eulan Dieldrin
(Cotton) 33.5 Kiering Scouring Bleaching	Sodium	Chloride Hypochlorite Polyphosphate Silicate Sulphate	Surfactants
33.6 Dyeing	Aluminium Chromium	Sulphite	Dyestuffs
(Synethetics) 33.7 Dyeing	Ammonium Copper Sodium	Bisulphate Chloride Hypochlorite Nitrate Polyphosphate	Benzoate Carboxy methyl cellulose Disperse dyes Formate Glyoxal Oxalate o-Phenyl phenol Peracetate Penthachlorophenol Phenol

Table II.2 (cont.)

Source	Inorganic		Organic
	Cations	Anions	
33.7 (cont.)			Salicylate Surfactants Trichlorobenzene
34. TIMBER			
34.1 Preservation	Arsenic Chromium Copper		Phenols
34.2 Preservation (in drying)			Pentachlorophenol
35. VISCOSE			
35.1 Rayon and cellophane etc.	Sodium Zinc	Sulphate Sulphide	Carbon disulphide Xanthate
36. WATER TREATMENT			
36.1 Softening and demineralisation	Calcium Sodium	Chloride	
36.2 Sewage treatment	Ammonia Sodium	Chloride Nitrate Nitrite Phosphate Sulphate	
36.3 Effluent treatment	Calcium	Chloride Sulphate	

Table II.3

POTENTIAL WATER POLLUTANTS

by Function

Function	Examples of specific pollutants		Sources (see key)
	Inorganic	Organic	
ANTI-OXIDANT		Methyl thiobenzothiazole Amines Phenols	24, 29 21 21
BLEACH	Borate Chloride Chlorite Hypochlorite Perborate Sulphite	Acetate Peracetate	4, 11, 27, 33
CHELATING AGENTS	Polyphosphate	EDTA Gluconate Heptonate NTA	11, 17, 33
COAGULANTS	Aluminium Iron	Chloride Sulphate	Acrylamide Polyacrylamide	36
COMPLEXING AGENT	Cyanide Fluoborate Fluosilicate		17
CONTRACEPTIVES		Ethinyloestradiol Mestranol Norathisterone Norgestrel and others	4 (22)
DYEING AUXILIARIES ETC.	Ammonium Sodium	Chloride Polyphosphate Sulphate	Acetate Benzoate Formate Glyoxal Oxalate o-Phenyl phenol Pentachlorophenol Peracelate Phenol Salicylate Surfactants (see below) Trichlorobenzene	33
FLOTATION AGENT		Xanthates (amyl, isopropyl)	18, 35
FLUX	Ammonium Zinc	Chloride		7
FUNGICIDES	Copper Iron Manganese Zinc	Dithiocarbamates	1 (25, 27, 34)
HARDENING SALTS	Barium Sodium Strontium	Cyanide		7 (17)
HERBICIDES	Chlorate	Atrazine 2, 4, D 2, 4, T	1
MONOMERS FOR PLASTICS MANUFACTURE		Acrylamide Acrylonitrile Formaldehyde Urea	10, 24, 36
OPTICAL BLEACH		p-Amino stilbene type Coumarin type Pyrazoline types Quinolene type Thiophene type	4, 33
PLATING SALTS	Cadmium Chromium Copper Lead Nickel Sodium Tin Zinc	Chloride Chromate Cyanide Fluoborate Fluoride Nitrate Sulphate	EDTA	17

Table II.3 (cont.)

Function	Examples of specific pollutants			Sources (see key)
	Inorganic		Organic	
PESTICIDES	Arsenic Lead		Carbamates DDT Dieldrin Nitrocresol Rotenone	1
PRESERVATIVES	Arsenic Chromium Copper		Penta chlorophenol Phenols	33, 34
SOLVENTS			n-Alkanes Benzene Acetone Isopropanol Methanol Methyl ethyl ketone Pyridine Toluene Xylene	5, 12, 13, 21, 22, 23
SURFACTANTS				

There is a wide variety of surfactants used industrially. The following are examples of the principal types:

- Anionic			Alkyl sulphates Dodecyl benzene sulphonate	4, 33
- Non-ionic			Nonyl phenol ethoxylate Alcohol ethoxylates	4, 11, 33
- Cationic			Cetyl pyridinium chloride Cetyl trimethylammonium bromide	11, 33
TANNING AGENTS	Aluminium Chromium	Borate Sulphate	Dimethylamine Formaldehyde Formate Phthalate	16

KEY TO THE TABLES

A. KEY TO CLASSES OF POLLUTANTS
(see directly Table II.1)

B. KEY TO SOURCES

1. Agriculture
2. Airports
3. Carbonization
4. Consumer products
5. Dyestuffs, intermediates and fine organic chemicals
6. Electronics
7. Engineering
8. Explosives
9. Fermentation
10. Fibreglass
11. Food processing
12. Glues and adhesives
13. Heavy organic chemicals and petrochemicals
14. Inorganic chemicals including fertilizer
15. Iron and steel
16. Leather
17. Metal Finishing

18. Mining
19. Non-ferrous metals
20. Paints and pigments
21. Petroleum
22. Pharmaceutical
23. Photographic
24. Plastics and synthetics
25. Power generation
26. Printing
27. Pulp and paper
28. Refuse disposal
29. Rubber
30. Soaps and detergents
31. Storage batteries
32. Storm water
33. Textiles
34. Timber
35. Viscose
36. Water treatment

C. KEY TO FUNCTIONS

1. Accelerator
2. Acid
3. Alkali
4. Alloy constituent
5. Anionic surfactant
6. Antibiotic
7. Anti-knock compound
8. Anti-oxidant
9. Anti-tarnishing agent
10. Bleach
11. Bronzing agent
12. Building material
13. By-product or waste product
14. Catalyst
15. Cationic surfactant
16. Chelating agent (see 20, 70)
17. Cleaner
18. Coagulant
19. Combustion product
20. Complexing agent (see 16, 70)
21. Contraceptive (oral)
22. Cutting fluid or component
23. De-icing fluid
24. Developer
25. Disinfectant
26. Dispersant
27. Dyestuff or auxiliary
28. Electrode material
29. Emulsifier
30. Etchant
31. Explosive
32. Fertilizer
33. Flocculant
34. Flotation agent
35. Flux
36. Fuel
37. Fungicide
38. Hardening salt

39. Herbicide
40. Inhibitor
41. Impurity
42. Insulator
43. Intermediate
44. Lubricant
45. Metabolic product
46. Metal
47. Monomer
48. Mordant
49. Mothproofing agent
50. Non-ionic surfactant
51. Odorant
52. Optical bleach
53. Ore
54. Oxidising agent
55. Paint or pigment
56. Pesticide
57. Phosphor
58. Photosensitiser
59. Plasticiser
60. Plastics
61. Plating solution
62. Polishing agent
63. Polymer
64. Preservative
65. Reducing agent
66. Retention aid
67. Rust inhibitor
68. Rust remover
69. Salt
70. Sequestering agent (see 16, 20)
71. Solvent
72. Stabilizer
73. Surfactant
74. Tanning agent
75. Unhairing agent

II.4 <u>HAZARD RATING</u>

A. OBJECTIVES OF THE SYSTEM

In the assessment and evaluation of water pollution a number of substances which cause particular concern are usually considered, but the number present may be too great for comprehensive action on all of them to be taken simultaneously. Thence priorities for action have to be set.

The purpose of the following section is to provide, for information purposes, a quantitative method of categorisation and ranking of water pollutants according to the hazard which they pose to man, wildlife and the environment expressed as a THP rating. This rating is based on measurable factors of Toxicity, Hazard and Persistence (THP) to give an overall numerical rating. This quantitative rating is supplemented by a magnification factor which is used when a compound is bio-accumulated, and by an indicator "c" for carcinogenic compounds.

Ideally a hazard rating should be simple, easy to understand and use, and not lead to inconsistencies in implied priorities, having regard to the quantities of the specific water pollutants used by industry, the concentrations at which they might occur, the widely differing toxicity (both acute and chronic) of different substances, and their direct or indirect modes of action.

Many experts have expressed the view that it is not possible to develop a single hazard rating covering adequately all harmful aspects of specific water pollutants and they have advised against the use of a hazard rating on the grounds that it might be taken out of context and misused. Others have recognised these difficulties, but have considered that it is necessary for decisions to be made and that some guidance on priorities is needed. Such guidance may not necessarily be related to remedial action but simply aimed at concentrating effort on a careful in-depth study of the problems resulting from the presence of a specific substance in water.

B. EXISTING HAZARD RATING SYSTEMS

Hazard rating systems have been developed which are applicable to the transport of hazardous chemicals. It would be advantageous if one of the systems, or a similar or compatible system could be used for specific water pollutants. It is therefore appropriate to examine the systems in some detail.

Several systems in common use are based on the United Nations Hazard Classification Numbers though names differ according to the country of origin and according to whether transport by sea, road or rail is being considered.

a) United Nations Hazard Classification Nos.

The United Nations Hazard Classification Nos. are used as a basis for the categorising of substances in "The Blue Book" dealing with the Carriage of Dangerous Goods in Ships (HMSO, London 1971), as a basis of the Kemler No. which is understood to be used in France, and as a basis for the hazard classification in the Hazchem system for the marking of road tankers conveying dangerous substances in the United Kingdom which was extended to rail transport in June 1976 (Annex II.1). The latter system supersedes the Voluntary Tanker Marking Scheme and the Transport Hazard Identification Scheme (THIS) in the United Kingdom.

The classification is as follows:

Class 1	– Explosives
Class 2	– Gases – compressed, liquefied or dissolved under pressure
Class 3	– Inflammable liquids
Class 4(a)	– Inflammable solids
Class 4(b)	– Inflammable solids or substances liable to spontaneous combustion
Class 4(c)	– Inflammable solids or substances which in contact with water emit inflammable gases
Class 5(a)	– Oxidising substances
Class 5(b)	– Organic peroxides
Class 6(a)	– Poisonous (toxic) substances
Class 6(b)	– Infectious substances
Class 7	– Radioactive substances
Class 8	– Corrosives
Class 9	– Miscellaneous dangerous substances, that is any other substance which experience has shown, or may show, to be of such a dangerous character that rules should apply to it

Class 10   - Dangerous chemicals in limited quantities

b) Kemler scheme

The Kemler scheme uses a composite number based on the United Nations Hazard Classification as an indication of the hazard. The first figure indicates primary hazard while the second and third figures if given indicate secondary hazards. Where the first and second figures are the same an intensification of the primary hazard is indicated, thus 33 means a highly inflammable liquid (flash point below 21°C); 66 indicates a very dangerous toxic substance; 88 means a very dangerous corrosive substance. Where the first two figures are 22 a refrigerated gas is indicated. The combination 42 indicates a solid which may give off a gas in contact with water. Where the hazard identification number is preceded by the letter X this indicates an absolute prohibition of the application of water to the product.

The Kemler system is described in Appendix B5 to the European Agreement concerning the Transport of Dangerous Goods by Road (HMSO, London 1974).

c) Hazchem scheme

This scheme at present relates to some 500 substances which are transported by road and rail tankers. The Hazchem Codes List No. 1 includes the United Nations No. for each substance, a Hazchem code, a Kemler number and other data. A specimen page is given in Annex II.1.

It is expected that legislation, making use of the Hazchem system mandatory for some 350 compounds, will be passed in the United Kingdom. The Health and Safety Executive published a consultative document in April 1977 entitled "Hazardous Substances (Conveyance by Road) Tank Labelling Regulations 1977 and Transport Hazard Information Rules".

The Chemical Industries Association in the United Kingdom has published a booklet entitled "Hazard Identification" which describes the scheme.

d) The European Economic Community

The hazard posed by chemical products in commerce and trade within the EEC has long been of concern to the Community. In the last decade and a half - since the Council Directive of 27th June, 1967 - some directives have been promulgated and proposed which include some elements of a hazard evaluation system. Although no comprehensive system has been published, reference may be made to the Official Journal, C34, dated 7th April, 1972, relating to labelling of products; and to the Official Journal, C260, dated 5th November, 1976, relating to the guidelines for evaluation of the chemical properties of substances put into trade. Further developments are being made in technical working groups and within the EEC.

e) GESAMP Hazard Profiles

A Group of Experts on the Scientific Aspect of Marine Pollution (GESAMP) has a Working Group on the Evaluation of the Hazards of Harmful Substances in the Marine Environment. This Group has developed a system of hazard profiles of selected substances. Unlike those in the systems mentioned above, the hazard profiles are not based on the United Nations Hazard Classification.

The profile considers five properties of each substance to which numerical values or symbols are assigned under column headings A, B, C, D and E (see Annex II.2 for specimen page). The legend to the hazard profiles is given in Table II.4.

A specimen page of such profiles is given in Annex II.2. Although the system is complex it does make a clear assessment of each potential hazard.

C. HAZARDOUS POLLUTING SUBSTANCES IN THE LOWER GREAT LAKES

A detailed study has been made by James F. Maclaren Limited, Environmental Consultants for Environment Canada on Hazardous Polluting Substances in the Lower Great Lakes. A list of hazardous materials was prepared in descending order of potential danger to the aquatic environment having regard to the toxicity to aquatic life, amounts used in commerce and industry, and mode of transport and storage.

To establish the list of hazardous materials, a methodology was developed for ranking the materials according to each criterion separately and for combining each into an overall rating system. The ranking of materials according to toxicity to aquatic life, E, incorporated the effects of lethal concentrations, solubility, volatility, biochemical half-life and physical state. For amounts used in commerce and industry the ranking, U, was based on production use, imports and exports of materials and extent of distribution or movement of materials. The ranking of materials according to

Table II.4

GESAMP SYSTEM

Column A - Bioaccumulation

+ Bioaccumulated and liable to produce a hazard to aquatic life or human health
O Not known to be significantly bioaccumulated
Z Short retention of the order of one week or less
T Liable to produce tainting of seafood

Column B - Damage to living resources

Ratings	TLm
4 Highly toxic	< 1 ppm
3 Moderately toxic	1 - 10 ppm
2 Slightly toxic	10 - 100 ppm
1 Practically non-toxic	≥ 100 - 1000 ppm
0 Non-hazardous	1000 ppm
BOD Problem caused primarily by high oxygen demand	
D Deposits liable to blanket the seafloor	

Column C - Hazard to human health, oral intake

Ratings	LD_{50}
4 Highly hazardous	< 5 mg/kg
3 Moderately hazardous	5-50 mg/kg
2 Slightly hazardous	50-500 mg/kg
1 Practically non-hazardous	500-5000 mg/kg
0 Non-hazardous	$\geq$ 5000 mg/kg

Column D - Hazard to human health, skin contact and inhalation (solution)

II Hazardous (solution)
I Slightly hazardous (solution)
0 Non-hazardous (solution)

Column E - Reduction of amenities

Ratings
xxx Highly objectionable because of persistency, smell or poisonous or irritant characteristics; beaches liable to be closed
xx Moderately objectionable because of the above characteristics, but short-term effects leading to temporary interference with use of beaches
x Slightly objectionable, no interference with use of beaches
0 No problem

All columns

Ratings in brackets, ( ), indicate insufficient data available to the Panel considering specific substances, hence extrapolation was required.

mode of transport and storage, R_c, considered four modes
of transport (ship, truck, rail and pipeline) and a uni-
form method of storage. Parameters considered for each
of the five conditions included quantity of material
transported or stored, accident potential, degree of
hazard, and accessibility of the spill to surface water
bodies. The procedure developed for combining these
independent rankings into an overall ranking system, OR,
was expressed by the following formula:

$$OR = E \times \sqrt{U \times R_c}$$

where E is the ecological effect rating, U is the overall
use distribution rating, and R_c is the combined accident
potential analysis. This is thought to be a thorough
and sound approach to assessing hazards associated with
production, use, transportation and storage of specific
substances. However, it does not necessarily indicate
appropriate priorities for action in respect of specific
water pollutants and is not generally applicable since
it is concerned specifically with the Lower Great Lakes.
It would be an enormous task to develop a similar system
for the major European rivers though the methodology
might well be followed in particular cases.

The ecological effect rating used in the formula
is a product of the assessment of toxicity based on
measurements of LC50 values and that of persistence,
approximately equal weight being given to each.

It is expressed by the equation

$$E = C^n (B + b)$$

where C = toxicity index based on a 4-day LC50 value
        (see Annex II.3.1)
      n = solubility moderator (see Annex II.3.2)
      B = persistence index (see Annex II.3.3)
      b = physical state correction (Annex II.3.4)

In taking account of different LC50 values a multiplier
is used as shown in Annex II.3.5. Annex II.3.6 gives
ratings corresponding to different categories of toxicity and persistence.

The overall use distribution rating is the product
of a use rating and a distribution rating.

The use rating is taken to be numerically equal to
the logarithm to the base 10 of the amount of substance
present (tonnes) divided by 10. A minimum value of 1 is
assigned. Therefore any quantity less than 100 tonnes
is given a use rating of 1.

The distribution rating is derived from a consideration of the frequency of movement of the substance under
consideration and is equal to the cumulative frequency
derived from Annex II.3.7.

The accident potential rating has regard to the
risks associated with transport of material and storage
of material expressed as an equation:

$$R_c = R_T + R_S$$

The risk or accident potential to the water environment for each mode of transport is represented by:

$$R_n = F_1 \times F_2 \times F_3 \times F_4$$

where R_n = the risk or accident potential associated with each of the n modes of
transport
F_1 = the quantity of material shipped by each
mode of transport
F_2 = the accident potential associated with
each mode of transport
F_3 = the size of the spill associated with
each mode of transport
F_4 = the portion of the spill which reached
the aquatic environment with each mode
of transport
R_s = the risk or accident potential associated with storage

Since each mode of transport has its own risk value
associated with it, the results of each is additive.
Therefore the combined risk for transportation, R_T is
given by

$$R_T = R_1 + R_2 + R_3 + R_4$$

To obtain a measure of the threat each hazardous
material poses to the water environment, and thus a ranking for hazardous materials moved by transportation
systems the following formula is used:

$$R = (Q_s \, H_s \, A_s \, S_s) + (Q_r \, H_r \, A_r \, S_r) +$$
$$(Q_t \, H_t \, A_t \, S_t) + (Q_P \, H_P \, A_P \, S_P)$$

where R_T = the total risk involved in transporting
a hazardous material
Q = category rating for quantity of material
shipped
H = the risk associated with shipment by
various means of transport

A = the fraction of spills gaining access
     to the water environment
S = the size factor of a spill from an acci-
     dent gaining access to water
Subscripts -
s = ship transport
r = rail transport
t = truck transport
p = pipeline transport

The risk associated with storage is considered to be the total quantity in storage and in ships, railway wagons, road trucks, and pipelines and is related to parameters Q H A and S described above.

Cursory examination of the overall hazard ratings indicates many apparent anomalies which presumably arise simply because of the different scale of production, use and transport of particular substances.

D. DEVELOPMENT OF A HAZARD RATING FOR SPECIFIC WATER POLLUTANTS

None of the hazard rating systems described above is ideal for categorising specific water pollutants. Nevertheless it would be desirable for any system adopted by OECD to draw on experience and methodology of other systems and to be compatible with them so that data for a specific substance obtained under one system can be incorporated for use in another. The rating proposed, which, it is suggested should be known as the THP rating in order to avoid direct reference to toxicity or hazard, draws on the methodology used in the GESAMP and Environment Canada systems. Except in respect of bioaccumulation which is difficult to quantity, it is proposed that the rating should be on a logarithmic scale related to the concentration at which the specific water pollutant causes damage in the aquatic environment, is a hazard to human health, or, if it is persistent, to the rate of degradation of the substance in the environment. The THP rating will be a composite of 3 numbers separated by a colon and followed by a symbol indicating those sub-stances which are bioaccumulated. The separate components of the overall rating are discussed in the following paragraphs.

a) Toxicity to the aquatic environment

This value is based on the lowest concentration at which the specific water pollutant has an adverse effect on the aquatic environment. This adverse effect may be of a widely differing nature for different substances.

For example, for cadmium it could relate to the effect of low concentrations on the eggs of salmon. For non-toxic paraffin oils it could relate to the concentration which causes an unsightly film on the surface of the water or interferes with the oxygen balance of the aquatic system. For chloride, it could relate to the concentration at which this ion would give rise to "meringue" dezincification of brass fittings on water distribution systems.

The rating proposed is numerically equal to the negative logarithm to the base 10 of the concentration expressed in g/l. This concentration could be the 4-day LC50 value where harm to fish is the appropriate criterion in which case the rating would be numerically similar to that in the GESAMP system, but it would be preferable to include an "application factor" (*).

Since one objective of the THP procedure is to sound a warning note, the most sensitive appropriate species should be chosen as indicator organism. This suggestion must sometimes be modified in practice. For example, selection of the most sensitive organism would also involve the risk that certain pesticides, for example, which are developed for reducing a specific noxious organism, but apart from this do as little harm to the environment as possible, are discriminated against and that this type of product development, which should be welcomed is in fact inhibited.

Clearly, long-term (chronic, sub-chronic) effects of pollutants are of considerable importance to man, either as a result of direct exposure by ingestion or as a result of ingesting bioaccumulated pollutants in food. At present the evaluation of these long-term risks is less than satisfactory and "so-called" safe values are only known for a few chemicals. To some extent, these values for known chemicals can be extrapolated to similar materials; however, in the absence of firm data the best available data from mutagenicity, teratogenicity, cyto-genicity and carcinogenicity assays must be used together with a small application factor.

The ratings corresponding to given concentrations are given below.

(*) Application factor =
Safe concentration for continuous exposure
—————————————
4 day LC50
(See reports of the United States National Technical Advisory Committee on Water Quality Criteria).

Concentration of specific water pollutant having an adverse effect on the aquatic environment	GESAMP rating	Toxicity value 'T' in THP rating	Quantity of specific pollutant significant in relation to 10^6 m^3 water
1 g/l 1 g/l 100 mg/l 10^{-1} g/l	0 Non-hazardous 1 Practically non-toxic	0 1	1000 t 100 t
10 mg/l 10^{-2} g/l	2 Slightly toxic	2	10 t
1 mg/l 10^{-3} g/l	3 Moderately toxic	3	1 t
100 µg/l 10^{-4} g/l	4 Highly toxic	4	100 kg
10 µg/l 10^{-5} g/l		5	10 kg
1 µg/l 10^{-6} g/l		6	1 kg
100 ng/l 10^{-7} g/l		7	100 g

b) <u>Hazard to human health by oral intake</u>

This rating would be similar but not identical to that in the GESAMP system. It is proposed that the numerical value of the rating should be equal to the negative logarithm of the LD50 value for oral intake expressed in g/kg. Examples are listed below:

Quantity of specific water pollutant giving rise to a hazard to health by direct oral intake	Hazard value 'H' in THP rating	Lethal Quantity (*) for 65 kg man
1000 mg/kg 1 g/kg	0	65 g
100 mg/kg 10^{-1} g/kg	1	6.5 g
10 mg/kg 10^{-2} g/kg	2	650 mg
1 mg/kg 10^{-3} g/kg	3	65 mg
100 µg/kg 10^{-4} g/kg	4	6.5 mg
10 µg/kg 10^{-5} g/kg	5	650 µg
1 µg/kg 10^{-6} g/kg	6	65 µg
100 ng/kg 10^{-7} g/kg	7	6.5 µg

*) Derived my multiplying the LD50 by 65.

c) <u>Persistence rating</u>

The proposed persistence rating is similar to that used in the report by James F. McClaren Ltd. for Environment Canada on the Lower Great Lakes. It is based on the probable half-life of the substance in the aquatic environment and is numerically equal to the logarithm of the half-life expressed in days with an upper limit of 3. This means that any substance with a half-life exceeding 1000 days is given a persistence rating of 3. Examples of the rating are given on following page:

Estimated half-life of specific water pollutant in aquatic environment	Persistence value 'P' in THP rating $\underline{/} = \log$ half-life (d)$\underline{7}$
1 d	0
10 d	1
100 d	2
1000 d	3 maximum value

d) Bioaccumulation rating

A substance which is bioaccumulated is by defini-
tion persistent and it seems appropriate to link the
bioaccumulation rating with that of persistence. It is
difficult to express bioaccumulation quantitatively.
The possibility of relating an assessment to a concentration factor was considered but rejected on the basis that
concentration of a substance can be from the aquatic
environment into many different parts of a living organism, or by means of a food chain. Thus, it would be
necessary to know whether to consider the whole organism
(e.g. a fish) or a specific organ of it. It is thought
to be adequate simply to draw attention to the fact that
a substance is bioaccumulated and to leave for special
consideration the extent to which bioaccumulation would
give rise to an increased hazard. In the proposed scheme
a plus symbol is used to indicate bioaccumulation.

e) Overall rating

The overall rating proposed is thus made up of
three separate numbers and in the case of bioaccumulated
substances a symbol. These relate to the toxicity to the
aquatic environment, T, the hazard to human health by
oral intake, H, and the persistence, P. It seems appropriate to call this the THP rating. In this way the
direct use of either toxicity or hazard is avoided and
it is thought that there will be less likelihood of the
rating being used out of context. The rating should be
easy to understand in that the greater the value of the
three component numbers, the greater the potential hazard arising from the presence of a particular concentration of substance. Relatively non-hazardous substances
would have a rating of 0:0:0 while a persistent substance
which has adverse effects at a concentration of 1 g/1
would have a rating beginning with 6 and ending with 3.

It is necessary to emphasize that it is not always
appropriate to relate the toxicity for the aquatic environment to a reported value of the 96-h LC50. For
example, the plant hormone 2,4 D is reported to be relatively non-toxic to fish (LC50 350 mg/1) yet it can cause
unpleasant taste in water at a concentration as low as

90

0.01 mg/l. Also aniline, though not acutely toxic to
fish and mammals, has adverse effects when constantly
administered at concentrations as low as 0.1 mg/l and
0.005 mg/kg. In using the THP rating, it is clearly
necessary to consider all the various hazards or problems
which might arise and select a suitable concentration or
dosage figure on which to base the rating.

As a pilot exercise, THP ratings, based on a narrow
interpretation of T have been derived for a number of
substances in the categorised list of specific water
pollutants, and these are shown in Annex II.4. If the
broad interpretation of T is used, the following results
are obtained:

Specific water pollutant	THP rating
Formaldehyde	3:1:1
Parathion	6:3:1
Dieldrin	6:2:3 +
Aniline	4:6:1
2,4DD	5:1:2
Pentachlorophenol	4:2:2
Orthocresol	5:0:1

This paper has suggested a basis for a THP rating.
Acquisition of basic data to propose confidently a rating
for each specific water pollutant will take considerable
time. It is suggested that it is better, at this stage,
to omit a rating where the data cannot be provided than
to put in a tentative one which may be misleading. However, where data are available in published literature,
but there is some doubt as to their reliability, it
would be possible to develop a rating and enclose this
within parentheses to indicate some element of
uncertainty.

It should be noted that the Hazard Rating scheme
developed her addresses mostly the question of acute
hazard to human health and longer-term hazards may not
be adequately covered. When the THP system is used,
every attempt should be made to obtain additional information on long-term health effects in order to obtain
a more meaningful assessment of the risk to human health.

f) Use of the THP system in practice

It must be stressed that great caution is needed in
using the THP rating system. Its use requires considerable knowledge, and because of the non-linear non-
arithmetic character of its component numbers of "scores"
more than simple addition is needed to establish a ranking among pollutants. The data require interpretation
and cannot be simply used to establish "norms" or even
less "standards".

In putting the THP system into practice certain other factors must be taken into account:

i) The field of concern has to be limited to keep it manageable, such as an industrial sector (e.g. tanning or pulp and paper) a special use (e.g. pesticides or detergents) or a group of compounds (e.g. chlorinated organic compounds or mercury and its compounds);

ii) It may often prove valuable to know the total amount of a compound released into the environment, and to investigate the distribution patterns for, say, individual amounts of substances used or discharged by an industrial branch. There may, alternatively, be a thorough study, for one or a few compounds, of the magnitude and directions of flow through society;

iii) In drawing the proper administrative conclusions from a rating procedure, it will be necessary to know about suitable substitutes (or alternative processes) for a compound. A compound which has a use, which is important to society, and which has no good substitutes is obviously more difficult to prohibit than a component of, say, a deodorant.

A VOLUNTARY SCHEME FOR THE MARKING OF TANK VEHICLES CONVEYING
DANGEROUS SUBSTANCES

HAZCHEM CODES, LIST No. 1

ALLOCATED BY THE JOINT COMMITTEE ON FIRE BRIGADE OPERATIONS AND CONFIRMED
BY THE HEALTH AND SAFETY EXECUTIVE

Substance	Tremcard Vol. & No.		Adr. Class No.		Kemler Number	United Nations Label	United Kingdom Regulations
Ammonia, anhydrous and solutions 50%	1	1	Id	5°	263	Toxic Gas	-
Butadiene	1	126	Id	6°	239	Inflammable Gas	-
Butane	1	176	Id	6°	23	Inflammable Gas	-
Butylene (Butene)	2	500	Id	6°	23	Inflammable Gas	-
Chlorine	1	2	Id	5°	266	Toxic Gas	-
Dimethylamine (Anhydrous)	1	73	Id	8°(a)	263	Inflammable Gas	-
Ethylchloride (Chloroethane)	3	616	Id	8°(a)	23	Inflammable Gas	-
Ethylene oxide	1	16	Id	8°(a)	236	Inflammable Gas	-
Hydrogen chloride (Anhydrous)	-	-	Id	10°	286	Toxic Gas	-
Isobutylene (Isobutene)	1	502	Id	6°	65	Inflammable Gas	-
Methylbromide (Bromomethane)	2	111	Id	8°(a)	263	Toxic Gas	-
Methylchloride (Chloromethane)	3	41	Id	8°(a)	236	Inflammable Gas	-
Nitrogen dioxide (Nitrogen Tetraxide) liquefied	1	109	Id	5°	265	Toxic Gas	-
Phosgene (Carbonyl Chloride)	1	107	Id	8°(a)	266	Toxic Gas	-
Propylene (Propene)	1	137	Id	6°	23	Inflammable Gas	-
Sulphur dioxide, liquefied	3	15	Id	5°	26	Toxic Gas	-
Vinylchloride, inhibited	3	150	Id	8°(a)	239	Inflammable Gas	-
Acetone	1	30	IIIa	5°	33	Inflammable Liquid	S.1.1971: 1040
Acrylonitrite, inhibited	1	61	IVa	2°(a)	633	Inflammable Liquid	S.1.1971: 1040
Amylacetates	1	581	IIIa	3°	30	Inflammable Liquid	-
Amylalcohols - primary and secondary	1	582	IIIa	3°	30	Inflammable Liquid	-
Amylene, normal (1-Pentene)	2	125	IIIa	1°(a)	33	Inflammable Liquid	Petroleum (Consolidation) Act 1928
Benzene (Benzol)	1	7	IIIa	1°(a)	33	Inflammable Liquid	"
n-Butanol (Butylalcohols)	1	583	IIIa	3°	30	Inflammable Liquid	-
Secondary Butanol (Secondary Butylalcohol)	1	583	IIIa	3°	30	Inflammable Liquid	S.1.1971: 1040
Tertiary Butanol (Tertiary Butylalcohol)	-	-	IIIa	5°	33	Inflammable Liquid	-
Butylacetate, normal	1	66	IIIa	3°	30	Inflammable Liquid	-
Secondary Butylacetate	2	518	IIIa	1°(a)	33	Inflammable Liquid	S.1.1971: 1040
Carbon disulphide (Carbon Bisulphate)	2	39	IIIa	1°(a)	336	Inflammable Liquid	S.1.1958: 257
Chlorobenzene (Monochlorobenzene)	2	90	IIIa	3°	30	Inflammable Liquid	-
Cyclohexane	1	103	IIIa	1°(a)	33	Inflammable Liquid	Petroleum (Consolidation) Act 1928
Cyclopentane	2	532	IIIa	1°(a)	33	Inflammable Liquid	"
Decahydronaphtalene (Decalin)	2	558	IIIa	3°	30	Inflammable Liquid	-
Dibutylethers (Butylethers)	2	584	IIIa	3°	-	Inflammable Liquid	-
Diethylether (Ethylether, Anaesthetic ether, Sulphuric ether)	1	72	IIIa	1°(a)	33	Inflammable Liquid	S.1.1971: 1040
2-6 dimethylheptan-4-one	2	554	IIIa	3°	-	Inflammable Liquid	-
Di-isopropylether	2	128	IIIa	1°(a)	33	Inflammable Liquid	-
Dimethylamine solution	-	-	IIIa	5°	-	Inflammable Liquid	S.1.1971: 1040
Dioxane	2	546	IIIa	5°	336	Inflammable Liquid	"

Substance	Tremcard Vol. & No.		Adr. Class No.	Kemler Number	United Nations Label	United Kingdom Regulations
Ethanol (Ethylal- cohol)	1	32	IIIa 5°	33	Inflammable Liquid	-
2-Ethoxyethyl- acetate (Ethylene glycol mono- ethyletheracetate)	2	459	IIIa 3°	30	Inflammable Liquid	-
Ethylacetate	1	76	IIIa 1°(a)	33	Inflammable Liquid	S.1.1971: 1040
Ethylbenzene	1	522	IIIa 1°(a)	33	Inflammable Liquid	Petroleum (Consolidation) Act 1928
Ethyl-n-butyrate	2	587	IIIa 1°(a)	-	Inflammable Liquid	
Ethylformate	2	537	IIIa 1°(a)	33	Inflammable Liquid	S.1.1971: 1040
Ethylmethylketone (Methylethyl ketone)	1	88	IIIa 1°(a)	33	Inflammable Liquid	"

SPECIMEN PAGE OF GESAMP HAZARD PROFILES
(see text for Legend)

Substances	Bioaccumulation	Damage to living resources	Hazard to human health		Reduction of amenities	Remarks
			Oral intake	Skin contact and inhalation (solution)		
	A	B	C	D	E	
Acetaldehyde	0	2	1	0	x	
Acetic acid	−	2	0	0	0	
Acetic anhydride	0	2	0	0	0	
Acetone	0	1		0	0	
Acetone cyanohydrin	0	4	3	II	xx	
Acetonitrile (Methyl cyanide)	0	0	1	0	0	
Acetyl Chloride	0	2	1	0	0	
Acrolein	T	4	3	I	xxx	
Acrylic acid	0	(2)	1	I	xx	? in Column B due to possible presence of unknown inhibitors.
Acrylic latex	0	?	0	0	xx	
Acrylonitrile	0	3	3	II	xxx	
Adiponitrile	0	1	3	I	x	
Aldrin	+	4	2	I	xxx	
Alkyl benzene sulfonate (straight chain)	0	2	1	0	0	
(branched chain)	0	3	1	0	0	
Allyl alcohol	0	3	2	0	xx	
Allyl chloride	0	2	2	0	xx	
Allyl isothiocyanate	0	(2)	2	II	xx	
Alum (15% solution)	0	1	0	0	0	
Alumina	0	D	0	0	0	
Aluminium phosphide	0	(3)	4	II	0	

Annex II.3

3.1 BASIC TOXICITY RATING

4-day LC_{50} Concentration	Rating Value (C)
Less than 0.1 mg/l	4
0.1 ppm to 100 mg/l	$3 - \log_{10} (LC_{50})$ (*)
Greater than 100 mg/l	1

*) LC_{50} is given in mg/l. The rating value, C, is evaluated to two (2) significant figures.

3.2 SOLUBILITY MODERATOR

$\dfrac{LC_{50}}{Solubility}$ Ratio (*)	n	Comment
Greater than 1,000	0	Minimal hazard. Solubility is sufficiently low. Neither acute nor chronic toxicity is expected to occur.
1,000 to 10	0.5	Solubility may depend on the chemical and physical characteristics of the aquatic environment. Ecological damage can occur during long-term exposure.
Less than 10	1.0	Solubility is greater than the accepted safe limits. Acute and/or chronic toxicity is likely to occur.

*) LC_{50} and solubility are expressed as mg/l. Solubility in distilled water at 20°C.

3.3 BASIC PERSISTENCE RATING

Probable Half-Life	Basic Persistence Index (B)
0 - 7 days	1.0
7 days - 1 year	1.5
Over 1 year	2.0

3.4 PHYSICAL STATE CORRECTION FOR THE PERSISTENCE INDEX

Physical Properties	Physical State Correction "b"
I. Soluble Materials; solubility >100 mg/l	0
II. Slightly soluble to insoluble materials; solubility <100 mg/l	
a) Gases: density <1 g/ml boiling point <0°C	-1.0
b) Light liquids and solids: density <1 g/ml boiling point >0°C	-0.5
c) Dense liquids: density >1 g/ml boiling point <300°	+0.5
d) Dense solids: density >1 g/ml boiling point >300°C	+1.0

3.5 MULTIPLICATION FACTORS FOR ESTIMATION EQUIVALENT TOXICITY VALUES

Toxicity Information (mg/l)	Multiplier for Equivalent Toxicity
1-day LC_{50}	0.25 x
2-day LC_{50}	0.50 x
4-day LC_{50}	1.00 x
5-day and 7-day LC_{50}	1.00 x
Tolerance Limit Value (TLV)	3.00 x
Others (including teratogenic, behavioural effect)	1.00 x

3.6 OVERALL ECOLOGICAL EFFECT RATING

Rating Value	Comment
0 - 1.5	Minimal hazards Low toxicity, Low persistence
1.5 - 4	Slight hazards, combination of low toxicity - high persistence or moderate toxicity - low persistence
4 - 6.5	Moderate hazards, Moderate toxicity - moderate to high persistence
6.5 - 9	Highly hazardous materials, Moderate to high toxicity, high persistence
9 - 12	Extreme hazards, high toxicity; very high persistence.

Annex II.4

HAZARD RATING SYSTEM FOR
SPECIFIC WATER POLLUTANTS

Having developed a system for hazard rating of specific pollutants, a trial was made of the system using a certain number of pollutants.

This annex gives trial results which on pragmatic grounds, are slightly modified from those originally foreseen. In particular, the original formula has been made more explicit by highlighting carcinogenicity by a "c" and the T values have been derived only from data on LC_{50} values for fish and the water flea (Daphne).

Class	Example	THP
Alcohols	Methanol	0:1:0
Aldehydes	Formaldehyde	2:1:0
	Glyoxal	1:1:1
Esters	Dibutyl phthalate	2:1:1
Hydrocarbons	n-Alkanes (hexane)	0:0:2
	Styrene	2:1:1
	Naphthalene	3:1:1
	3,4-Benzofluoranthene	7:7c:3
	3,4-Benzopyrene	7:7c:3
Ketones	Acetone	0:0:0
	Methyl isobutyl ketone	1:1:0
Organo-halogens	Ethyl chloride	0:0:0
	Vinyl chloride	1:3c:0
	1,2-Dichloroethylene	2:2c:3
	Bis (2-chloroisopropyl) ether	4:1c:2
	Chloroalkanes	3:1:2
	DDT	6:3c:3+
	Dieldrin	6:3c:3+
	Eulan	2:0:0
	Pentachlorophenol	5:2:2
	Polychlorinated biphenyl	6:6c:3+
	Atrazine	3:1:3
	2,3,6-Trichlorobenzoic acid	1:2:3
	2,4-Dichlorophenoxyacetic acid	2:4:1
	1,2,4-Trichlorobenzene	4:2:1
	Hexachlorobenzene	6:5:3+
Organo-phosphorus	Parathion	7:5:3+
Organo-metal	Bis (tributyltin) oxide	5:1:3
Organo-nitrogen	Acrylamide	1:1:1
	Acrylonitrile	2:1:1
	Carbamates	3:3c:1
	Dimethylamine	2:1:0
	EDTA	1:0:3
	NTA	1:1:0
	Urea	0:1:1
	Aniline	4:1c:1
	Nitrobenzene	2:1:0
	Nitrophenols	3:2:0
	Nitrocresols	3:2:0
	Pyridine	3:1:1
	Morpholine	1:1:1
	p-aminostilbene based dyes	3:2:2
	Cetyltrimethylammonium bromide	4:2:3
Organo-sulphur	Carbon disulphide	3:2:2
	Sulphurised hydrocarbons	2:2:2
	Xanthates	2:2:2
	Thiophene	3:2:2
	Dodecylmercaptan	3:3:1
	Alkyl benzenesulphonates	3:1:1-3
	2-Mercaptobenzothiazole	2:1:1
Phenols	Phenol	2:2:0
	Lignin	0:0:3
	o-Phenylphenol	3:1:0
	Cresols	3:1:0
Salts	Formate	2:0:0
	Acetate	1:0:0
	Propionate	1:0:0
	Stearate	1:0:0
	Naphthenate	2:1:1
	Oleate	1:0:0
	Palmitate	1:0:0
	Oxalate	1:0:1
	Tartrate	0:0:0
	Peracetate	2:1:1
	Heptonate	2:1:0
	Gluconate	0:0:0
	Benzoate	1:0:0
	Salicylate	1:1:1
	di-n-butylphthalate	3:1:3+
Steroids	Cholesterol	1:1:3+
	Ethinyloestradiol	6:7:3+

Class	Example	THP
Steroids (contd.)	Mestranol	6:7:3+
	Norethisterone	5:6:3+
	Norgestrel	6:7:3+
Metals and Metalloids	Aluminium	3:0:3
	Antimony	2:1:3
	Arsenic	4:3c:3
	Barium	2:3:3
	Beryllium	4:6c:3+
	Cadmium	6:5c:3+
	Calcium	0:0:3
	Chromium	3:1:3
	Cobalt	3:2c:3
	Copper	4:1:3
	Iron	2:2:3
	Lead	3:4:3+
	Manganese	2:2:3+
	Magnesium	0:0:3
	Mercury	6:6:3+
	Molybdenum	1:1:3
	Nickel	3:2:3
	Potassium	0:0:3
	Silver	5:3:3
	Sodium	0:0:3
	Strontium	1:0:3
	Tin	2:2:3
	Titanium	2:2:3
	Uranium	2:3c:3+
	Vanadium	2:2:3
	Zinc	3:0:3
Cations	Ammonium	2:0:0
Anions	Borate	1:0:3
	Chloride	0:0:3
	Chlorine	5:3:0
	Chlorate	2:0:2
	Cyanide	4:4:1
	Ferricyanide	2:2:1
	Ferrocyanide	2:2:1
	Fluoborate	2:2:3
	Fluoride	2:2:3
	Fluosilicate	1:1:3
	Nitrate	0:2:0
	Nitrite	3:2:0
	Phosphate	1:0:3
	Polyphosphate	1:0:3
	Polysulphide	5:3:1
	Silicate	1:0:3
	Sulphamate	1:1:0
	Sulphate	0:0:3
	Sulphide	5:3:1
	Sulphite	1:1:0
	Thiocyanate	1:2:1
	Thiosulphate	1:1:0
Solids	Asbestos	1:6c:3

c = carcinogenic
+ = bioaccumulated
LC_{50} values to fish and daphne have been used as a basis of calculation of T.

PART III

ORIGIN, OCCURRENCE AND CONTROL OF
SPECIFIC POLLUTANTS IN POTABLE WATER

III.1. <u>INTRODUCTION</u>

Most water sources in Member countries receive some form of treatment prior to distribution as potable supplies. Originally the methods of purification have been developed with the primary objectives of controlling the spread of micro-organisms and of reducing the total quantities of suspended materials. The control of individual specific pollutants has only rarely been considered, with some interest being occasionally directed at a handful of conspicuous coloured, odoriferous or taste-producing substances. More recently, however, increasing realisation of the complexity of water pollution problems, with thousands of specific pollutants already known*, has underlined the need for a thorough re-evaluation of existing purification procedures. The need for such a study is further emphasized by the recognition that certain of the identified pollutants are now known, or suspected, to be carcinogens or to have other serious adverse effects. Of special relevance are recent epidemiological studies, noting possible correlations between increased incidences of heart disease, bladder, stomach and other cancers and the consumption of water from particular surface sources.

A primary cause for concern is the apparent deterioration in many of the raw waters currently used as potable sources. This deterioration reflects, on the one hand the worldwide increase in the use of chemicals and on the other hand the continually rising demands placed on limited available water resources. The increase in water demand, moreover, is often very uneven with disproportionately heavy demands being placed on a few readily accessible sources near urban areas. One consequence of this is the increasing indirect reuse of water in areas of high demand, e.g. the Thames. As a result, dry weather water flows of certain rivers may contain high proportions of waste water.

In these circumstances, adequate treatment becomes more necessary, but also more difficult and problematic. In this section the extent to which current potable

---

* Consumption of plastics, for example, has risen ten-fold since 1950(1). Consumption of many inorganics shows scarcely less dramatic growth. Thus production of refined cadmium since 1950 amounts to no less than 80 per cent of all of the metal used since its isolation in 1817(2).

water treatment procedures are able to control the increased contamination by specific pollutants is considered. It is recognised that the preparation of potable water from low quality sources, as has been increasingly frequent over the past decades, should be positively discouraged and gradually abandoned.

Finally, it should be remembered that even where general pollution control measures are effectively enforced, raw waters will still often contain a baseload of pollutants from: uncontrolled industrial and domestic discharges; diffuse or non-point sources; residual and "bio-fractory" pollutants not effectively removed by waste water treatment plants; and from all types of accidents which may affect water quality.

At the request of Member countries, a special study has been carried out on the problem of organochlorinated compounds present in potable water resulting from chlorination processes currently used. This particular issue is dealt with in detail in Part IV and is therefore only summarised here.

III.2. OCCURRENCE OF SPECIFIC POLLUTANTS
IN RAW WATERS AND TRENDS IN POLLUTION

The problems arising from various inorganic contaminants have been known for some time and considerable data on concentration levels are currently available. Information about most organic compounds, however, is much more limited. Relatively recent advances in analytical techniques for these compounds have greatly increased the flow of relevant information, particularly for those 10 per cent of organics which are volatile. Nevertheless, data on the larger nonvolatile fraction are more limited. It is clear that this gap in present knowledge may seriously hinder the formulation of effective measures. Published reports already list hundreds of individual pollutants and, it is likely that the total reaches several thousand. The large majority of these compounds are organic.

A. PRESENT SITUATION

Summary lists of inorganic and organic pollutants are given in Tables III.1 and III.2 respectively. It will be seen that reported concentrations for inorganic pollutants range as high as 140 μg/l for lead and may even reach 5,000 μg/l for highly soluble boron. Considerable differences in inorganic pollutant levels between continents are also evident in Table III.I. Amongst the organics listed in Table III.2 are several reported to have occurred in concentrations exceeding 200 μg/l. For instance, for the following hazardous pollutants, the concentrations given are maximum levels reported in drinking water.

3.4 benzofluoranthene	-	1.1 μg/l
PCBs	-	up to 3 μg/l
Vinyl chloride	-	up to 10 μg/l
Bis (2 chloroethylether)	-	up to 0.42 μg/l
Chloroform	-	up to 366 μg/l
Carbon tetrachloride	-	up to 5 μg/l

Groundwater is generally better protected against any rapid inflow of pollution. Also, extended periods of underground residence enable biological mineralisation and physico-chemical equilibration with the components of the surrounding soil and rock.

106

Table III.1

INORGANIC SPECIFIC POLLUTANTS IN SURFACE WATERS

Parameter	Microgram/Litre					
	Observed values in waters of the United States 1962/1967			Observed values in waters of the European countries 1969/1973		
	Max.	Mean*	No. of observations	Max.	Mean	No. of observations
Antimony	N.D.	N.D.	N.D.	1.0	0.5	2
Aluminium	2800	74	456	N.D.	N.D.	N.D.
Arsenic	340	64	87	27	6.0	200
Barium	340	43	1568	N.D.	N.D.	N.D.
Beryllium	1.2	0.19	85	N.D.	N.D.	N.D.
Bismuth	N.D.	N.D.	N.D.	350	71	21
Boron	5000	100	1546	520	82	67
Cadmium	120	9.5	40	19	2.1	336
Chromium	110	9.7	386	960	19	206
Cobalt	48	17	44	21	5.4	157
Copper	280	15	1173	130	12	259
Iron	4600	52	1192	3800	630	191
Lead	140	23	305	130	24	194
Manganese	3200	58	810	650	190	240
Mercury	N.D.			5.7	0.64	159
Molybdenum	1500	68	506	6	1	57
Nickel	130	19	256	110	15	249
Selenium	N.D.	N.D.	N.D.	20	3.2	98
Silver	38	2.6	104	8	3	6
Strontium	340	43	1571	500	330	3
Vanadium	300	40	54	70	5.6	62
Zinc	1200	64	1207	570	140	141

* Values below detecting limits excluded from calculation of mean.

(Sources: E.P.A., Washington and EEC, Brussels).

N.D. = No Data

Table III.2.

MAXIMUM CONCENTRATIONS OF SOME ORGANIC MICROPOLLUTANTS
IN SURFACE WATERS

Substance	Surface water	Country	Concentration (μg/l)
Acetic Acid	Ohio River	United States	25
Aldrin	Rhine River	Netherlands	0.15
2-Aminotoluene	Rhine River	Netherlands	1
Aniline	Rhine River	Netherlands	9.8
Benzenehexachloride	Rhine River	Netherlands	0.81
Benzidine	Rhine River	Netherlands	0.4
3,4-Benzofluoran-thene	Meuse River	Netherlands	1.1
3,4-Benzopyrene	Meuse River	Netherlands	1
Benzylether	Rhine River	Netherlands	1
n-sec-Butylacetate	IJssellake	Netherlands	5
tert. Butylmer-captan	Ohio River	United States	4.6
n-Butyric Acid	Ohio River	United States	0.18
Carbophenothion	Essex River	United Kingdom	8
Chlordane	Red River	United States	0.075
2-Chloroaniline	Rhine River	Netherlands	2.2
Chlorobenzene	Rhine River	Netherlands	5
bis-Chloroisopropy-lether	Rhine River	Netherlands	25
4(4-Chloromethyl-phenoxy)-Butyric Acid	Essex River	United Kingdom	0.15
2(4-Chloro-2-Methyl-phenoxy) Propionic Acid	Essex River	United Kingdom	1.3
-Chlorotoluene	Rhine River	Netherlands	20
Cholesterol	Thames River	United Kingdom	1.1
Cyclohexanol	Rhine River	Netherlands	10
p.p. DDE	Kent River	United Kingdom	5.5
p.p. DDT	Rhine River	Netherlands	0.03
Diazinon	Essex River	United Kingdom	0.016
Dibutylphthalate	Rhine River	Netherlands	10
3,4-Dichoroaniline	Rhine River	Netherlands	2.9
o-Dichlorobenzene	Rhine River	Netherlands	60
2,4-Dichlorophenol	Ohio River	United States	6.6
2,4-Dichlorophenoxy Acetic Acid	Yakima River	United States	0.33
Dieldrin	Kent River	United Kingdom	0.059
N,N'-Diethylamine	Rhine River	Netherlands	10
Di-Isobutylcarbinol	Kanawa River	United States	27
Dimethylaniline	Rhine River	Netherlands	1
Diphenylether	Rhine River	Netherlands	0.2
alpha-Endosulphan	Rhine River	Netherlands	0.06
Endrin	Rhine River	Netherlands	0.02
Fluoranthene	Rhine River	Netherlands	3.4
Geosmine	Seine River	France	3
Heptachlor	Rhine River	Netherlands	0.04

Table III.2 (Contd.)

Substance	Surface water	Country	Concen-tration (μg/l)
Hexachlorobenzene	Rhine River	Netherlands	0.22
Hexachlorobutadiene	Rhine River	Netherlands	50
p-Isopropyldiphenyl-amine	Rhine River	Netherlands	1
Malathion	Essex River	United Kingdom	0.3
Methylethylpyridine	Rhine River	Netherlands	1
2-Methylisoborneol	Wabash River	United States	0.1
Methylstearate	Thames River	United Kingdom	2
2-Methylthiobenz-thiazole	Rhine River	Netherlands	200
Naphthalene	Rhine River	Netherlands	3
Nicotinic Acid	Knahwa River	United States	3
Nitrobenzene	Rhine River	Netherlands	20
p-Nitrochloro-o-benzene	Mississippi River	United States	37
o-Nitrophenol	Scheldt River	Netherlands	1.3
o-Nitrotoluene	Rhine River	Netherlands	20
p-Nonylphenol	Rhine River	Netherlands	10
Phenylmethylcar-binol	Kanawha River	United States	17
Phenylphenol	Rhine River	Netherlands	0.1
PCBs	Rhine River	Netherlands	0.4
Tetrachloroiso-propylether	Rhine River	Netherlands	0.6
o-Toluidine	Rhine River	Netherlands	2
Toxaphene	Clayton Lake	United Kingdom	28
Trichlorobenzene	Rhine River	Netherlands	5
2,3,6-Trichloro-benzoic Acid	Essex River	United Kingdom	1
Trichlorophenol	Kent River	United Kingdom	40
2,4,5-Trichloro-phenoxy Acetic Acid	Zurich Lake	Switzerland	0.8
1,3,5-Trimethyl-benzene	Rhine River	Netherlands	10

Note: Some of these data are from single observations, others from multiple observations.

(Source: Water Research Centre, Stevenage, U.K.; National Institute for Public Health; and National Institute for Water Supply, Netherlands).

These data were presented at the "International Working Meeting on Health Effects relating to Direct and Indirect Reuse of Waste Water for Human Consumption" which was organised by the World Health Organisation (WHO) International Reference Centre for Community Water Supply (1975).

Groundwater contamination is, however, occurring increasingly frequently from many polluting sources which include: agricultural activities (use of fertilisers, pesticides), industrial discharges to deep wells, leakage from fuel tanks and waste dumps, and a range of contaminants from roads (salt, metals, oil etc.). These potential hazards are particularly serious because most existing groundwater sources are not routinely analysed for such pollutants.

The normal chemical composition of groundwater reflects the regional geological characteristics. Thus groundwater shows considerable variation between sources both for inorganic substances and for humic and fulvic acid composition. Significant levels of radioactive substances* and/or physiologically active components including arsenic, selenium, fluoride and lithium may also occasionally be present. Table III.3 provides details of a number of groundwaters used for potable supply. It is notable that levels of ammonium ion are generally high and that a wide variety of elements such as zinc, barium and arsenic may also be present in considerable concentrations.

A particularly serious problem is the rapid build-up of nitrate levels, especially in groundwaters. Many European countries are now severely affected, the main cause being the ever increasing use of nitrogenous fertilisers and the resulting massive leaching to the aquifers. The drinking water standard of 45 mg/l (NO_3^- has already been exceeded in many supplies, and the phenomenon is expected to increase in the future, both in extent and in intensity. (The reader should refer to the OECD study on this subject: "Water Contamination by Nitrate as a Result of Intensive Fertiliser Use".)

In contrast to the often high concentrations of inorganic substances, concentrations of organic compounds are generally moderate in groundwaters. Thus, a recent survey of the impact of organic biocides suggests that there is no evidence, as yet, of a widespread contamination(5). Evidence from "accidental" pollution incidents suggests, however, that the problem faced is not a short-term one but is more likely to be a slow, steady process, the effects of which may only become evident after many years.

Other well-recorded incidents of groundwater contamination include that by solid waste tips near Ockenburg (Netherlands) where water from wells now shows considerable concentrations of lead, copper, zinc and cadmium. Similarly, coal gas residues buried during the 1920s near Ames (Iowa), although removed in 1954, have

---

* Drinking water levels of up to 25 pCi/litre of Radium 226 have been noted in the United Stated(3).

Table III.3.

ANALYSES OF GROUNDWATERS
(MEDICINAL AND SPA WATERS)

Parameter and unit of measurement		Number of observations	Max. conc.	Standard
Ammonium	mgNH$_4$/l	13	2	0.05*
Arsenic	μg/l	13	190	50
Barium	μg/l	2	330	100*
Beryllium	μg/l	11	27	
Boron	μg/l	4	2000	
Cadmium	μg/l	8	1	10
Calcium	mg/l	15	460	200
Chloride	mg/l	15	390	600
Chromium	μg/l	11	31	50*
Cobalt	μg/l	10	1	
Copper	μg/l	11	30	1500
Iron	μg/l	15	4500	1000
Lead	μg/l	11	18	100
Lithium	μg/l	14	3750	
Magnesium	mg/l	15	127	150
Manganese	μg/l	14	1000	500
Mercury	μg/l	5	0.3	1
Nickel	μg/l	11	110	50*
Nitrate	mgNO$_3$/l	12	9.2	45
Orthophos-phate	mgPO$_4$/l	7	1.3	
Potassium	mg/l	9	90	12*
Sodium	mg/l	10	1800	100*
Strontium	μg/l	4	980	400
Sulphate	mgSO$_4$/l	15	1182	
Titanium	μg/l	1		
Zinc	μg/l	11	60	1500

WHO recommended drinking water standard 1975 is quoted.
Those marked by an * are the adopted E.E.C. drinking
water standards 1975 (this code also includes proposed
levels for other parameters covered by the WHO
recommendations listed above).

contributed considerable quantities of coal tar materials to the city water supply. Not only taste and odour problems have resulted from this contamination but individual analyses have detected acenapthalene (19.3 ug/l) and a variety of other toxic polynuclear hydrocarbons.

River water contamination - Table III.4 summarises typical values for 15 important parameters in rivers with varying degrees of pollution. Fairly high levels of ammoniacal nitrogen, chloride, fluoride and sulphate can occur whatever the pollution situation. Considerable increases in nitrate, ortho phosphate, hardness, salinity, iron, zinc, anionic detergents and mineral oil are likely to be noted as pollution increases. Not unexpectedly, there is a tendency for levels of specific pollutants to be higher downstream.

Because of the larger volumes of water present, lake water tends to change in composition more slowly and thus pollutant levels generally rise only after several years. For this reason, pollution abatement programmes are likely to be equally slow acting. In this respect, lakes tend to occupy an intermediate position between rivers and groundwater sources. Particular difficulties have been encountered in many lakes following the build-up of phosphates and nitrates which has lead to increased eutrophication, and in turn to considerable problems for drinking water supply.

B. TRENDS IN THE QUALITY OF WATER SOURCES

Generally, changes in water quality follow from changes in the type and intensity of human activities in the area. It is expected, therefore, that in the absence of controls, the more densely populated and the more industrialised a country becomes, the larger its pollution problem is likely to be. This principle has been established, using nations as the environmental units, by Goldberg and Bertine, who proposed the ratio of GNP to area as an approximate measure of national potential pollution.

The OECD Member countries which have the highest GNP/AREA ratio according to this concept and which have a population of more than one million are listed in Table III.5. This table also shows the ratio between GNP and mean annual precipitation which may be more meaningful in national water pollution problems. Although national ratios of this type are of value in overall planning, it is generally more appropriate to consider smaller environmental units, such as river basins or lakes, for the description of quantity trends in specified water sources.

112

Table III.4.

WATER QUALITY DATA ON POLLUTED RIVERS

YEARLY AVERAGE VALUES IN RIVER WATER

Parameter	Unit	naturally/*slightly artificially polluted				moderately* artificially polluted				strongly* artificially polluted			
		Min.	Max.	Mean	n**	Min.	Max.	Mean	n**	Min.	Max.	Mean	n**
Ammonia	mg/l	0.1	1.1	0.6	2	0.01	3.1	0.52	8	0.03	2.7	0.51	8
Chloride	mg/l	1.0	138	29	21	1.7	100	21	35	2.4	300	61	28
Fluoride	mg/l	0.01	1.8	0.40	11	0.05	1.3	0.29	22	0.03	1.1	0.37	13
Nitrate	mg/l	0.05	8.0	1.8	16	0.01	23	4.0	28	0.1	25	9.0	25
O-Phosphate	mg/l	0.01	2.5	0.34	13	0.005	1.3	0.25	28	0.1	9.0	1.4	24
Sulphate	mg/l	14	240	120	15	3.8	100	46	11	25	160	88	7
Total Hardness (measured as CaCO3)	mg/l	11	260	100	9	17	320	120	19	14	400	210	8
Iron	mg/l	0.005	1.3	0.29	17	0.008	5.8	0.89	28	0.05	4.6	1.2	17
Manganese	mg/l	0.01	0.04	0.02	3	0.01	0.32	0.10	15	0.01	0.28	0.10	7
Zinc	mg/l	0.021	0.04	0.036	3	0.005	0.30	0.045	13	0.03	0.44	0.26	3
Anionic Detergent	mg/l	0.02	0.05	0.035	20	0.005	0.15	0.061	3	0.02	0.63	0.19	4
Gamma-BHC	µg/l					0.005	0.01	0.01	1	0.007	0.1	0.04	8
pp'DDT	µg/l					0.005	0.02	0.007	3	0.002	0.05	0.016	4
Dieldrin	µg/l					0.005		0.012	2	0.001	0.011	0.006	6
Oil	mg/l						1.0	0.4	3	0.5	3.8	1.6	4

These data are based on unpublished information, obtained from the WHO International Reference Centre for Community Water Supply and involve yearly average concentration values for a total number of 83 river sites in the world. River sites belonging, at this moment, to the naturally/slightly artificially polluted class are, e.g. the Blue Nile at Khartoum, Sudan, the Mekong at Chau-Doc, South Vietnam and the Glomma at Kongsvinger, Norway.

Within the moderately artificially polluted class river sites can be found the Gota Alv at Göteborg, Sweden, the Tisza at Szeged, Hungary, and the Mississippi at East St. Louis, United States (Illinois). Strongly artificially polluted rivers are, e.g. the Sambre at Namur, Belgium, the Trent at Nottingham, England and the Rhine at Lobith, the Netherlands.
* See Table III.9. for details of the water category definitions.   ** = No. of observations.

Table III.5

RANKINGS OF NINE OECD MEMBER COUNTRIES
WITH THE HIGHEST NATIONAL
POTENTIAL POLLUTION RATIOS (1970)

Rank	Country	GNP/Area ($10^9/year/km^2)	GNP Precipitation ($/m^3)
3	Netherlands	0.75	1.1
4	Belgium	0.56	0.69
5	Japan	0.54	0.30
6	Germany (Fed. Rep.)	0.52	0.65
7	Switzerland	0.41	0.23
9	United Kingdom	0.39	0.40
10	Italy	0.24	0.24
11	Denmark	0.23	0.32
12	France	0.18	0.23

Examples outside OECD countries

Rank	Country	GNP/Area	GNP Precipitation
1	Hong Kong	2.5	1.3
2	Singapore	2.3	0.93
8	Puerto Rico	0.40	0.40
12	Germany (Dem. Rep.)	0.19	0.32
13	Israel	0.19	0.48
15	Lebanon	0.16	0.26

The key parameter for prediction of the potential trend of water quality in lakes might be GNP*/unit of volume of water or, for rivers, GNP*/unit of flow in yearly average discharge (m^3/sec). The general relationship between GNP/flow ratio and several water quality parameters is given in Annex I which illustrates that, in general, the increasing load of waste products from human settlements can first be detected in rivers by increasing levels of coliforms followed by increases in phosphate and nitrate. Later, rises in the levels of chloride, fluoride, iron, zinc, manganese and copper are likely. Although local variations will occur, it is likely that socio-economic yardsticks of this type can be valuable for predicting trends.

---

* Average value for the appropriate drainage area of the water body under consideration.

Data on changes in specific pollutant levels over long periods are rarely available. One time series of considerable interest, however, is that relating to the Rhine where chloride has been noted to increase linearly from about 15 mg/l in 1875 to about 200 mg/l in 1975(8). A second important trend established with reference to the Rhine concerns the increasing percentage of organic residues which are likely to be non-degradable (bio-refractory). Thus, as biological waste water treatment becomes more common, it is inevitable that the relative amount of non-biodegradable substances will increase. For example, on the Rhine, it has been estimated that the level of non-degradable organic compounds is con-siderably increasing and may soon represent a significant percentage of total organic carbon present(9). Potable water treatment will then require special purification stages in addition to biological treatment.

Another time series relevant to prediction of pollu-tion trends concerns levels of inorganic contamination found in historical ice samples. In areas of Greenland, for example, annual ice layers are identifiable back to approximately 1200. These layers reflect the levels of contamination annually received as dust, snow and rain. In the 670 years up to 1870 levels of certain metallic components rose by approximately 70 per cent. By the 1950s however an increase of 210 per cent had occurred and by the 1960s an increase of 460 per cent. It is likely that surface water sources, especially those in Western Europe and North America, will have received increased quantities of pollutants at least at the same rate as the ice samples, and probably at a far higher rate.

III.3. ORIGIN OF SPECIFIC POLLUTANTS IN RAW WATERS

Pollutants in raw waters used for preparation of potable water come from a variety of sources among which two broad categories are: point sources of pollution and non-point (or diffuse) sources.

A. POINT SOURCES: INDUSTRIAL AND MUNICIPAL EFFLUENTS

These sources of pollution, derived directly from manufacture, use and disposal of goods, represent the largest contribution to the total. Industrial effluents usually contain the largest quantities and greatest variety of pollutants; however, municipal effluents which not only include domestic and commercial waste waters but also those of small industries plus some urban run-off also provide a considerable pollutant load. These sources are described in greater detail in I.3.

An important proportion of effluents (variable from country to country) still receives no treatment at all. Furthermore, even treated effluent cannot be considered as pure water. Bulk pollution (BOD and suspended solids) may be eliminated only to 50 per cent or 60 per cent on a 365 day average. Moreover, many specific pollutants are poorly removed in conventional treatment processes.

B. NON-POINT OR DIFFUSE SOURCES

The significance of diffuse sources is not yet quantified but because these sources are not directly treatable, they add to the background or residual pollution which must be removed from raw waters at potable water treatment plants. In some river basins, diffuse sources may contribute up to half the total pollution load. It is known, for example, that the run-off from urban/industrial areas is a significant contributor of heavy metals (lead, zinc, cadmium, nickel, chromium etc.) and other undesirable substances such as asbestos, detergents and hydrocarbons(18). All these materials will, of course, eventually be washed off by rain and largely transferred to the sewer system. The zinc content of galvanising (which also often contains a significant

116

proportion of cadmium) is likely to be completely dissolved within 20 years in moist atmospheric conditions
and since over one quarter of total zinc consumption is
used annually for galvanising(2), large quantities of it
and of cadmium will become potential water pollutants.
Agriculture is a main contributor of nitrates in underground and surface waters, and this causes concern in relation to potable water quality, and eutrophication.
Agriculture is also a major source of biocides, many of
which are toxic and persistent.

a) Run-off from Urban/Industrial Areas and Activities

This includes in particular:

i) run-off from urban and industrial areas, road
systems, airports, harbours etc. (not connected
to a sewer network and not treated);
ii) drainage from domestic, industrial or mining
refuse dumps (including sludges), infiltration
from underground oil tanks and uncontrolled
"black wells" etc;
iii) fall-out from airborne pollutants mostly generated in urban/industrial areas. A large part
of them fall to the ground within a short
distance of the source (large and medium-sized
particles); while others may be airborne for
much longer distances. Furthermore, acid compounds (in gaseous, liquid or particulate
phase) cause considerable corrosion and thus
secondary pollution.

b) Agricultural Sources in General

These include:

i) Spraying of biocides of all sorts (herbicides,
fungicides, defoliants, insecticides, rodenti-
cides etc.) and often in significant
quantities(16);
ii) Application of fertilisers (nitrates);
iii) Intensive animal husbandry, ensilage of crops;
iv) Other sources: run-off from agricultural land
and forests; natural or man-made erosion.

c) Accidents

Because of their unpredictability as regards time
and place, accidental discharges of pollutants have to be
considered as diffuse sources. However, three types of
accident situation are recognisable:

i) During transport and transfer of products by
sea, rail, road and pipeline;
ii) During storage of products;
iii) Sudden discharges from manufacturing processes
resulting from neglect, equipment and malfunc-
tion or emergencies.

Accidents can contribute a wide range of pollutants to
the water environment, many of which will have substantial effects e.g. "fish kills" which occur from time to
time in industrialised rivers.

III.4. POLLUTANTS GENERATED DURING TREATMENT AND DISTRIBUTION OF WATER

A category of pollutants now causing increasing concern are those which arise during actual "treatment" and distribution. These pollutants are of three main types:

a) substances picked up by contact with the structure of the treatment plant and subsequent distribution network;

b) substances present as impurities in treatment reagents;

c) substances formed by reactions between purifying and disinfecting agents and the initial pollutants. These reactions may occur at the treatment plant and also during passage through the distribution pipework.

In the past, awareness of these problems has been limited, partly because of the common practice of taking water samples at the treatment plant exit rather than at the actual point of use (i.e. domestic tap), and partly because better analytical techniques now enable detection of pollutants at low concentrations (μg/l).

A. POLLUTANTS PICKED UP BY CONTACT

During the treatment and distribution processes, water comes into contact with a wide variety of structures. In consequence, quantities of adventitious materials, both dissolved and as fine particulates are acquired and may be distributed to the consumer. Asbestos is one solid material which may be picked up in this way, whilst dissolved pollutants include metals and a variety of organics. The origins and significance of each of these important contaminants are discussed below.

Asbestos - levels in raw water are usually below 10 million fibres/litre but levels of 100 million fibres/litre are known, particularly in urban areas and drinking water levels of 1.9 million fibres have been reported in some cities(28). In many cases the fibre content mainly results from industrial operations or asbestos deposits, but the use of asbestos cement piping in distribution systems may also contribute to the asbestos content in

119

drinking water, particularly in areas where the water supply is soft and aggressive. In a preliminary study of drinking water supplies in the United States concerning leaching or erosion of chrysotile, levels up to 10,000 fibres/litre were detected(29). Methods for asbestos determination in water are not yet standardised nor even reliable. This lack of a good analytical method is a serious handicap and its solution is a matter of concern and urgency.

Although there is no epidemiological evidence that ingestion of asbestos fibres through drinking water is a health hazard(30)(31), it is considered that, in view of possible health effects, the level of asbestos fibres in potable water should be maintained as low as possible by the best available technology; there is no known "safe" level for asbestos.

Metals - all common metals are soluble, particularly in soft corrosive water supplies. Uptake from pipeworks, storage tanks and tap bodies is consequently universal. Although they are generally slight, higher metal concentrations can occur where water storage times are long, in the first flow from the tap, in new pipework lacking protective deposits and where waters are soft. Among the potentially toxic metals which may be acquired in this way are lead, copper and cadmium.

Lead - levels in raw water are generally low and a level of 0.1 mg/l is only exceeded in a few final water supplies. However, where lead pipes are used for connecting households to the mains or where water is drawn from lead-lined storage tanks, higher concentrations may arise. Thus, from a survey of over 3,000 houses carried out in the United Kingdom in 1975, it was estimated that about 1.6 per cent of first draw tap water samples exceeded the WHO recommended lead level of 0.3 mg/l and that 4.3 per cent of all households exceeded 0.1 mg/l in random daytime samples(22). Recent estimates suggest that the adult intake of lead from water and food is likely to be about equal where water lead levels average 0.1 mg/l.

Copper - The levels involved /up to 280 µg/l in raw waters and up to several mg in tap supplies(3)7 are likely to be of concern only to the small number of individuals who suffer from Wilson's disease.

Cadmium - Concentrations of this metal are generally much lower than those of lead and rarely exceed a few micrograms/l. (Only 0.1 per cent of United States water supplies in the survey already mentioned exceeded 10 µg/l(3). However, tap levels may be much higher(23). The main sources of cadmium in drinking water are the low-grade zinc used in galvanising equipment and storage tanks, solder (up to 50 per cent cadmium) and hardened copper pipework (1 per cent cadmium). This has been a

serious problem in Scandinavia, especially with hot water.
Cadmium used as a stabiliser in plastic piping may also
cause problems.

Organic Pick-up - As in metallic pipework, plastic
structures may contribute significant adventitious
material to the contained water. Examples include vinyl
chloride from low grade PVC with levels of up to 5.5 µg/l
being noted(24), and various polyaromatic hydrocarbons
probably derived from coal tar **pipe** linings(25). Other
recent data have shown migration of some organic compounds such as gases and liquids of low molecular weight
from soil through plastic pipes.

B. CONTAMINATION BY USE OF IMPURE REAGENTS

There is increasing use of physico-chemical treatments with different chemicals. Lime is the most traditional chemical used but modern treatment processes uti-
lize salts of iron and aluminium; flocculation adjuncts;
activated silicates, alginates, synthetic polyelectrolytes; neutralisation agents such as sodium hydroxide
and bicarbonate; and disinfection/sterilisation agents
like chlorine. Many of these reagents - particularly if
they are of "technical" or lower grade purity - are
sources of contamination. Table III.6 summarises some
recent data on the type of contaminants found in typical
water treatment reagents.

A practical approach to controlling pollution from
the reagents is to impose purity standards for them.
Such standards should ensure that the normal amount of
reagent used contributes less than 10 per cent of the
standard level of the particular contaminant in treated
water. For reagents not yet fully certified for use in
potable water treatment, special steps are required to
avoid any health risk. For example, the amount of monomer in polyelectrolytes must be controlled and
standardised.

Table III.6

POLLUTANTS ADDED WITH SOME WATER TREATMENT REAGENTS OF TECHNICAL GRADE PURITY

Reagent	Typical major contaminants	Other contaminants
Sodium Chlorate	Chlorate	
Aluminium Sulphate	Iron, Arsenic, Lead	Manganese, Cadmium
		Cadmium, Nickel
Iron Sulphate	Manganese	Mercury, Lead
		Cadmium, Arsenic
Sodium Aluminate	Iron, Arsenic, Lead	
Iron Chloride	Manganese, Arsenic	Copper, Nickel
		Zinc, Chromium
Sodium Hydroxide	Iron, Aluminium	
Sodium Carbonate	Ammonia, Iron	
Sulphuric Acid	Iron, Arsenic, Lead	
Hydrochloric Acid	Iron, Chlorinated-organics	
Activated Carbon	Copper, Zinc, Lead, Arsenic	
Ammonium Chloride	-	Nickel, Lead Copper
Sodium Bisulphate	-	Selenium, Lead Zinc
Calcium Hydroxide	Mercury	
Chlorine	Carbon Tetrachloride Other Halogens	Various organo-halogens

Source: (reference 20)

C. CONTAMINATION FROM PRODUCTS FORMED DURING PURIFICATION PROCESSES

Recently a major area of concern has been the compounds generated following the addition of treatment agents and the fear that some of these compounds may present health hazards. There is some concern about the various by-products generated by the oxidants (such as the organohalogenated compounds). Yields of by-products are likely to be significant when raw waters with high organic content are used. Unfortunately these raw waters are being utilised increasingly.

a) <u>Chlorine</u> - First adopted as a potable water disinfectant in 1908(3), this reagent provides an effective means of micro-organism control, it is cheap, and consequently is now almost universally used. In addition to being a disinfectant, chlorine is also widely used during the treatment itself as an oxidant in degrading <u>organic</u> substances and <u>ammonia</u> (break-point chlorination). Current procedures involve addition of chlorine as a gas, a liquid or as hypochlorites. Chlorination may either preceed or follow filtration; pre-chlorination requires larger quantities of chlorine and also produces larger quantities of organochlorines. Chlorine is also added as a terminal stage in treatment to provide a persistent residual disinfection (bacteria-static) in the distribution network. Moreover, it is frequently applied to raw waters before transportation.

i) <u>Quantities used</u>. Originally chlorine in moderate doses was used only in post-treatment as a final disinfectant in order to obtain a small residual concentration. Under these circumstances most of the chlorine is present as chloramines (which are also a disinfectant but require longer contact time). Since this time, conditions of chlorine use have considerably changed in many treatment plants and it is now used in high doses, and at the beginning of treatment. This procedure leads to increased organochlorine formation, and once formed these are largely resistant to further treatment.

Current practice with average intake waters involves the addition of 1 mg/l or more of chlorine with the aim of establishing a "free" chlorine residual of 0.1 - 0.4 mg/l. Where high initial levels of organics are involved, higher doses are being used, which, in exceptional cases, may reach 40 mg/l. This type of treatment is increasingly common for potable water sources. Such treatment practices for raw waters containing significant levels of organic substances generally lead to the formation of considerable quantities of organochlorines and it is not, therefore, unusual to find these still present in drinking water at the end of treatment.

The reasons for concern about the by-products of chlorine treatment are twofold - the hazardous character of certain compounds formed, and the total quantities produced.

ii) <u>Molecules formed in drinking water</u> - Some 280 halogenated compounds have already been found in drinking water and others will doubtless be identified over the next few years. A wide variety of chemical categories are represented. They include molecules related to PCBs, insecticides, herbicides, halogenated ethers, halogenated paraffins, halogenated phenols and many others. Of the individual compounds, chloroform is the commonest being found virtually throught the

123

United States and in many of the European waters studied
(24); concentrations up to 366 μg/l have been reported(3).
Table III.7 shows some recent illustrative data on formation of halogenated compounds as a result of potable water
chlorination. About 33 per cent of the 400 or so known
refractories are chlorinated(27).

iii) Quantification - The second cause for concern
regarding chlorinated residues centres on the massive
quantities currently being generated and the widespread
distribution of the products. Thus, when 129 United
States supplies were recently sampled "virtually all"
showed chlorinated organics. It has been estimated that
up to about 3 per cent of applied chlorine may be re-
tained in organic form. Most estimates suggest that
3-4 per cent of total chlorine production (approximately
22 million tonnes in 1974 for OECD countries) is used for
water treatment. The scale of pollution by chlorinated
organics is perhaps best seen in comparison to the quantities involved with other hazardous pollutants. For the
same period, mercury production amounted to 9,000 tonnes,
cadmium(2) to 17,000 tonnes and PCBs to 46,000 tonnes
/OECD only (27)7.

Table III.7

THE EFFECT OF CHLORINATION ON THE OCCURRENCE

OF SOME HALOGENATED COMPOUNDS

IN DIFFERENT TAP WATERS

Compound	Concentrations (μg/l)		
	No chlori-nation	Disinfec-tion only	Break-point
Chloroform	<0.01-2.0	<0.1-10	25-60
Bromodichloro Methane	<0.01-0.9	<0.01-10	15-55
Dibromochloro Methane	<0.01-0.1	0.01-5	3-10
Dichloroiodo Methane	<0.01	<0.01-0.3	0.01-0.3
Bromochloroiodo Methane	<0.01	<0.01-0.03	<0.01-0.3
Bromoform	<0.01	<0.01-1.0	3-10
1-1 Dichloroacetone	<0.005	<0.005	0.1-1.0
Trichloronitro Methane	<0.01	<0.01-3.0	<0.01-3.0

Source: (reference 19)

b) Ozone

First used in 1906 and now employed in more than
1,000 plants, notably in France, ozone is an even more
powerful oxidising and disinfecting agent than chlorine

124

but has no residual effect in the distribution network. Ozone has various advantages, particularly in relation to taste and odour but, being a strong oxidant, it might also lead to production of undesirable by-products such as ketonic compounds. Little is known about the formation of these substances but prudence is desirable and ozone or other oxidants should preferably be used at minimum dosage at the end of the treatment. Ozone does not remove ammonia and in some cases may hinder the subsequent biodegradation of certain substances. According to local conditions, a light application of a disinfectant with a residual effect (e.g. chlorine, chlorine dioxide) may be required when ozone is used. Ozone is more expensive than chlorine, but the cost is quite acceptable for a moderate final application.

c) Chlorine Dioxide

Chlorine dioxide has been used satisfactorily for the last two decades in potable water treatment in certain Member countries (Switzerland, France, United Kingdom, United States etc.) and in quite a large number of plants. It was originally used instead of chlorine to improve the taste and odour of potable water. Its disinfectant and residual effects are greater than those of chlorine and it can be used in smaller doses. Its cost is somewhat higher than that of chlorine but this is partially compensated for by the lighter application required. The increased interest being attached to chlorine dioxide in several Member countries is because as a final disinfectant, it has qualities at least equivalent to those of chlorine and does not result in the formation of halomethanes in potable water to the same extent. It might however, give rise to chlorite formation. Comparative trials on this reagent have been launched in a number of Member countries. From the initial experience now gained in full-scale application in potable water plants, the following provisional results can be stated:

- chlorine dioxide should only be used as a final disinfectant in light doses (0.1-0.3 mg/l);
- for other purposes, such as oxidisation of ammonia or organic substances, other techniques (e.g. biological filtration) should be used.

Compared with chlorine, its use requires more careful adaptation, since with existing techniques it has to be prepared at the point of use. This is an important issue

as, unless it is prepared in the optimum way* small
quantities of chlorine or sodium chlorite may also be
formed and then chlorine dioxide may lose some of its
specific advantages.

d) Activated Carbon

Treatment with activated carbon and particularly
filtration through granulated carbon has developed considerably in recent years. The main characteristic of
activated carbon is its ability to adsorb substances
without "exchanging" them for others (e.g. ion-exchange
resins). However, unless it is well managed and controlled, filtration on activated carbon can itself cause
noticeable deterioration in water quality for two main
reasons:

- desorption or elution can occur if the carbon is
  not of sufficient quality, and when it becomes
  oversaturated after a certain period of use;
- careful surveillance is needed when regenerated
  carbons are put back into service because, in the
  case of inadequate regeneration, undesirable substances may be liberated.

e) Water softening

In a number of Member countries, domestic softeners
have been indiscriminately installed on a large scale by
commercial firms. It is now recognised that water
softening should be restricted to use with water heaters
and washing machines. However, in many cases all the
water consumed in homes, including that for drinking and
cooking, is softened. Softening has two negative effects
from the health viewpoint: magnesium, and calcium ions,
which are useful elements of human diet, are removed; and
secondly, they are replaced by sodium ions which generally are already in excess in human diet and are associated
with cardiovascular and other effects. This subject has
caused some controversy over the past few years, but the
various studies performed sufficiently confirm the need
for concern. Moreover, such uncontrolled softening may
make the water corrosive and thus lead to dissolution of
hazardous metals from piping. Proliferation of micro-
organisms on the ion exchanger resins of softeners has

---

* The recommended method is to prepare the chorine
dioxide by reaction of sodium chlorite with a solution of
chlorine (5 g/l).

also often been noted. Strict regulations on water
softening should be taken where this has not been done
already, and the public should be better informed of the
dangers of such equipment. In addition, neutralisation
of acid waters in potable water treatment plants is
often carried out on a large scale with caustic soda because dosage is so easy. For health reasons, this should
be discouraged and lime should be used instead.

III.5. SPECIFIC POLLUTANTS IN DRINKING WATER

Although there are many surveys of specific pollutants in raw waters, there is much less information on their levels in drinking water as supplied to the consumer. The World Health Organisation is a major body concerned with the quality of drinking water, and their International Reference Centre for Community Water Supply (the Hague) has many data on specific pollutant levels.

Tables III.8 and III.9 summarise recent data on specific pollutants in drinking water which illustrate the current problem.

Table III.8

SPECIFIC POLLUTANTS IN DRINKING WATER

Chemical class of compound	No. of times found in drinking water	Type most frequently found	No. of times found in drinking water
Alcohols	46	– Aliphatic Alcohols	38
Aldehydes	47	– Aliphatic Aldehydes	37
Alkane Hydrocarbons	96	– Normal Alkane Hydr.	76
Alkene Hydrocarbons	14	– Alkene Hyd.	12
Amines	13	– Aliphatic	8
Benzenoid Hydrocarbons	99	– Alkybenzenes	77
Carbohydrates	3	– (not specified)	3
Carboxylic Acids	21	– Aliphatic Long-chain (C > 7)	10
Esters	78	– Phthalates	47
Esters & Heterocyclic O2 Compounds	40	– Aliphatics	24
Aliphatics	470	– Aliphatic Chlorides	257
Halogenated Aromatics	125	– Chlorinated Benzenes Alkyl Benzenes	82
Ketones	98	– Aliphatic	72
Nitro-compounds	9	– Aromatic	6
Miscellaneous Nitrogen Compounds	27	– Heterocyclic	14
Pesticides & Herbicides	59	– Polycyclic Halogenated	24
Phenols & Naphthols	21	– Alkylphenols	12
Phosphorus Compounds	6	– Phosphates	6
Polynuclear Aromatic Hydrocarbons	71		
Sulphur Compounds	34	– Compounds with S-O Bonds	

Source: (reference 21)

Table III.9

ORGANIC SPECIFIC POLLUTANTS IN TAPWATER

Compound	No. of times found in 20 tapwater samples	Highest detected concentration µg/l
Toluene	20	0.3
Xylenes	19	0.1
C_3-Benzenes	19	1.0
Decanes	18	0.3
Ethyl Benzene	17	0.03
Fluoranthene	16	0.05
Nonanes	15	0.3
Naphthalene	14	0.1
Dibthyl Phthalate	17	0.1
1-1 Dimethylisobutane	13	0.3
Methyl Isobutyrate	13	1.0
Chloroform	16	60
Tetrachloromethane	15	0.7
Bromodichloromethane	11	55
Dibromochloromethane	12	20
Bromoform	7	10
Trichloroethene	9	9
Trichloronitromethane	3	3
Bis (2-Chloroisopropyl) ether	8	3
Octanol (Isomers)	8	3
Dichloroidomethane	4	1
1-1 Dichloroacetone	3	1
2-Cyclohexen-1-one	1	1
Hexyl Butyrate	1	1

Source: (reference 19)

130

III.6. EFFECTS OF SPECIFIC POLLUTANTS
IN DRINKING WATER INCLUDING HEALTH EFFECTS

A. GENERAL PRINCIPLES

The setting of standards for drinking water quality depends, of course, on the effects which pollutants in the water have on consumers. Adverse effects are sometimes classified as either health hazards or nuisance effects but these tend to overlap. This section highlights some of the better known and documented effects of specific pollutants.

Some specific pollutants may have important powers of water coloration, taste or odour formation. Examples include dyes - particularly the more conspicuous colours such as yellows and reds and various oily substances capable of forming surface films. In relation to taste and odour, the problems are often detected by the high sensitivity of human sensory organs. Certain chlorophenols and chlorobenzenes, for example, are readily detectable by taste and odour at levels well below the sensitivity of modern analytical instruments.

Eutrophied waters may have a very unpleasant taste (especially in summer), which is extremely difficult to eliminate. This "musty" taste results especially from metabolites (geosmine, isoborneol, mucidone) largely due to cyanophyceal and associated actinomycetes; degradation of organic matter in anaerobic conditions also contributes. The high content of soluble organic matter found in eutrophied waters and not eliminated by potable water treatment, moreover, may reactivate the growth of microorganisms in distribution networks and lead to the formation of organochlorines when chlorination takes place. An imbalance in the water (excess acidity, for example) may not only create a variety of nuisance effects for the user and the distribution networks, but also lead indirectly to certain health effects from, for example, metal dissolution.

It is recognised that unpleasant tasting drinking water indicates that poor quality raw waters have been used, and implies that a variety of specific pollutants may be present even if they are not "individually" detectable by taste. In fact the public intuitively

131

reacts in this way and to some extent avoids drinking tap water when it tastes bad. This reaction is particularly noticeable in certain OECD countries and apart from some obvious social impacts (on the family budget for instance) it may also have significant health effects which have often been denounced by the medical profession because it leads to:

- regular consumption of mineral bottled waters, many of which are medicinal waters (not "potable" according to national standards due to their high mineral content) and should not be absorbed over long periods;
- daily consumption (even by young children of a large variety of commercial beverages which, from a health viewpoint, are far less desirable than water (beverages containing chemical flavourings, sweetners, colourings, preservatives, high sugar content or alcohol).
- finally, it has been noted that a deficit in daily human intake of water may result from a poor tasting water supply and this may affect health (e.g. kidneys).

B. ACUTE TOXICITY

In view of the concentrations involved, acute toxicity due to chemical components in water is rare in OECD countries. An example is methaemoglobinaemia, a disease affecting infants who have ingested water with a high nitrate content. The disease involves the formation of nitrites in the intestinal tract and affects the blood haemoglobine (blue babies). The instances are generally geographically limited and more frequent in summer when both nitrate content of water and incidence of gastroenteritis are higher. The health hazard of nitrates/nitrosamines is discussed below.

C. LONGER-TERM EFFECTS

Certain specific pollutants present particular problems because they are only poorly excreted by man and may thus be cumulative in their effects. In these circumstances, consumption of water, even with very low levels of contamination, can lead to significant build-up over a period of years. This possibility is increased by the tendency of consumers to draw their water supply from a limited number of sources even over long periods of time. Food, in contrast, is typically now drawn from a wide variety of sources and any individual high levels of

contamination thus tend to be compensated for. Examples
of cumulative hazards from water include lead and cadmium
poisoning(7-13).

D. CARCINOGENIC AND SIMILAR EFFECTS

By far the greatest cause for concern in connection
with specific pollutants centres on those substances which
are suspected or proven carcinogens. They are of special
concern because the quantities of carcinogens required to
produce an irreversible effect may be well within the
quantitative range which might be acquired from water over
a period of time. Some of the pollutants identified as
being present in drinking water are recognised carcinogens.

Current evaluation of this hazard is largely based
on extrapolation from laboratory animal feeding studies;
some 6,000 substances, in all, have now been tested for
carcinogenesis in this way. A second source of information is mutagenicity testing; of some 118 waterborne
compounds tested to date, no fewer than 12 have proved
positive. Such studies differ, however, from the water
situation in the concentrations used and the length of
exposure. A further limitation is that synergistic (or
antagonistic) effects such as might occur in a typical
mixture of water pollutants cannot be readily duplicated
by such studies. Table III.10 lists a number of known
or suspected carcinogens found in water. Many of these
substances are chlorinated molecules which may already
have been in the raw water, but come principally from the
chlorination of water supplies.' Finally, a number of
compounds are known to act as co-carcinogens, among which
are phenols and petroleum derivatives.

A final source of information is based on epidemiological studies comparing disease incidence between areas
with differing water characteristics. First initiated to
investigate the historically high level of cancer mortality in New Orleans, further studies have recently been
done in Ohio, New Jersey and elsewhere. A number of
such studies have now been completed in different countries and all appear to show an association between
organic contaminants and increasing cancer rates, although
the magnitude of the effect is not yet clear(15). In
connection with the time scale of the effect, it should be
borne in mind that the latency period for cancers is
about 15 - 20 years (range 5-30). Any current increased
mortality would therefore largely reflect the levels of
contamination existing between 1950 and 1970.

One group of carcinogens which are causing particular
concern are the nitrosamines, some of which rank among the
most carcinogenic compounds known, and many of which are

Table III.10

SOME ORGANIC SUBSTANCES FOUND IN DRINKING WATER AND KNOWN OR SUSPECTED TO BE CARCINOGENIC

	µg/l maximum values reported	Reference number
Vinyl chloride	10	2
3.4 Benzopyrene	< 0.005	10
Dieldrin	8	2
BHC	0.01	2
Bis (2 chloro ethyl) ether	0.42	2
Chlorodane	0.1	2
3.4 Benzofluoranthene	0.444	10
CTC	10	4
PCBs (Penta chloro biphenyl (Tetra chloro biphenyl (Tri chloro biphenyl	3	2
Benzophenone	1	10
Heptachlor	0.01	10
Chloroform	366	2
Tri chloro ethylene	0.5	2
Aldrin	< .01	10
Nitrites (Nitrosamine precursors)	30	8
Atrazine	5.1	10
Endrin	0.08	2
Hexachloro benzene	0.19	10
Hexachloro butadiene	0.27	10
Hexachloro ethane	4.4	1
1.4 Di chloro phenol	0.5	1
Bromo dichloro methane	116	1
Bromoform	92	1
Chloro benzene	5.6	1
Di bromo chloro methane	100	1
1.2 di chloro ethane	6	1
Tetra chloro ethane	0.11	10
Methylene chloride	1.6	1

also hepato-toxins. The concern is because nitrosamines may be formed in the intestinal tract from normal dietary proteins in the presence of excess nitrites and nitrates. As has already been noted, nitrate pollution is increasing and nitrate removal is, in practice, difficult. The subsequent increased nitrate intake by populations, from water and other sources, may be a problem of considerable significance which cannot be neglected any longer. Corrective measures are required, such as careful control of nitrates in drinking water, and regulations on fertiliser use and agricultural practices.

E. HEALTH EFFECTS OF SOME INORGANIC POLLUTANTS

There is a vast literature on environmental and health effects of inorganic specific pollutants, and much of this section draws on the 1977 report "Drinking Water and Health" of the National Academy of Sciences, Washington, D.C.

a) <u>Cadmium</u> - Food is the primary source of cadmium intake for man, and drinking water provides only about 5 per cent of the total daily intake. Cadmium ingestion at a higher level than 100 µg/day (in the presence of other dietary constituents) caused itai-itai disease in Japan. Major toxic effects in humans are on the kidney. Some animal studies have shown carcinogenic and teratogenic effects but the dose/response relationships are not yet known. Cadmium has also been implicated as a factor in hypertension. It should be noted that there is a relationship between zinc, cadmium and calcium concentrations, and that zinc and calcium protect, to some extent, against cadmium toxity. (<u>Zinc</u>, contrary to cadmium with which it is frequently associated, is a relatively non-toxic element. Concentrations of about 30 mg/l give a strong astringent taste and milky colour to water, and at higher levels some acute adverse effects have been reported. There are no known chronic adverse effects of low level long-term ingestion of zinc).

b) <u>Chromium</u> - Concentrations of chromium found in natural waters are limited by the low solubility of trivalent chromium oxides. The trivalent form is converted to the hexavalent during chlorination, and it then passes through other treatment processes (in a soluble anionic form). High doses of hexavalent chromium are toxic, producing erosion of the gastro-intestinal tract and kidney lesions.

c) <u>Cobalt</u> - Cobalt has only been observed in natural waters in trace amounts. Food is the major source of cobalt, and normally less than one per cent of total daily intake is from water. Excessive cobalt intake may cause heart failure. Acute toxic effects in animals only occur, however, in massive doses. Chronic toxicity has been reported in children taking cobalt to correct anaemia.

d) <u>Copper</u> - Because the average daily intake of copper is about 2.5 mg/day, and because water can contain up to 1 mg/l copper, the daily intake from water can exceed that from food. The general health hazard from copper intake at a few mg/l appears to be small. However, a few people with a copper metabolism disorder - Wilson's disease - can be adversely affected by ingestion of even trace amounts of copper.

e) <u>Lead</u> - Under normal circumstances only about 10 per cent of the daily intake of lead is from potable water, and the natural lead content of surface waters is generally small. There are no known beneficial effects of lead in humans or animals. Although acute lead poisoning is rare, chronic lead toxicity is severe and occurs with even a low daily intake (<1 mg) because lead accumulates in bone and tissue. Children, especially in cities, are a special risk group for lead toxicity because their dangerous absorption threshold for lead from water intake can be reached much more readily than that of adults. The major chronic adverse effects of lead are produced in the haematopoietic system, central and peripheral nervous system and the kidneys. The synthesis of haemoglobin is the biochemical reaction which is thought to be the most sensitive to lead.

f) <u>Manganese</u> - Manganese is found less frequently and usually at lower concentrations than iron. Even ingestion of manganese in moderate excess of 3 - 7 mg/day is not considered harmful, although in some unusual geological situations very high manganese concentrations can occur and cause the disease manganism.

g) <u>Mercury</u> - Mercury is a comparatively rare element and most of its inorganic compounds are relatively insoluble and can exist in solution only in extremely small concentrations under natural conditions. However, industrial use of mercury has resulted in serious environmental pollution. The adverse health effects on people occupationally exposed to mercury and mercury compounds have long been recognised, but pollution of the general environment is of recent origin. In nature, inorganic mercury in bottom sediments of water bodies can be transformed biochemically to methylmercury or other organic compounds which are injurious, and which readily enter the food chain and can concentrate up to 3,000 times in fish. Symptoms of mercury intoxication appear at daily mercury intakes of about 250 to 900 µg.

h) <u>Molybdenum</u> - Soluble molybdate ions are present in trace concentrations in many surface waters, and in normal circumstances the intake from water is a very minor part of the total daily intake of humans. Some finished water supplies are, however, reported to have as much as 1 mg/l molybdenum and there is little coherent knowledge of adverse health effects of these orders of concentrations. In fact, molybdenum poisoning has rarely been observed in humans (although it has been implicated in gout in Armenia and in a bone disease in India); however, the cause/effect relationship is still to be established. It may be noted that copper tends to offset molybdenum toxicity.

i) <u>Nickel</u> - Nickel is not removed from water by normal treatment processes but is generally present at low concentrations in surface and drinking waters. It is thought to be moderately absorbed from food and water; concern has arisen primarily because inhalation of nickel compounds increases the risk of lung and nasal-cavity cancers. Moreover, it has recently been pointed out that liver cancer might be associated with nickel absorption; the long-term effects of this metal are not yet fully understood.

j) There are other <u>inorganic</u> ions which possibly have harmful health effects when present in potable water. These include nitrate, nitrite, sulphate, cyanide and among the metals sodium, selenium, arsenic and boron. For detailed information reference should be made to(3) and to WHO publications.

III.7 STANDARDS FOR SPECIFIC POLLUTANTS
IN DRINKING WATER

It is generally recognised that standards required
for drinking water should be more comprehensive and
detailed when the quality of the raw water from which it
is prepared is lower. Not only should the standards be
more detailed but the monitoring more frequent and
rigorous.

As a result of the use of increasingly contaminated
raw water and of the development of analytical methods
for pollutants at very low concentrations, a large number
of standards have been worked out recently for drinking
water and raw water used for making drinking water.
Some standards for reagents and materials to be used in
potable water management have been developed but a great
deal of work still remains.

Although knowledge of maximum allowable concentrations of specific pollutants is needed, it is inevitable
in practice to aim at overall parameters of water quality
like turbidity, total organic carbon, total organic
chlorine and the taste and odour index. If stringent
values for these overall parameters are met, it is also
probable that individual specific pollutants will be
present only in relatively small quantities. Stringent
values for these overall parameters are thus very important, to protect the consumer against the presence of unknown, possibly toxic, compounds in polluted raw waters.

The directive of the Commission of European Communities defines standard treatment methods for transforming
surface water of categories A1, A2 and A3 into drinking
water. The lowest acceptable is A3 and its treatment
includes intensive physical and chemical treatment,
extended biological treatment and disinfection (e.g.
chlorination). Surface water falling short of A3 standards must not be used for abstraction of drinking water,
or only in exceptional circumstances where extra purification processes are used. Because there is some uncertainty about the long-term health effects of many specific pollutants and as these pollutants are widely present
in raw waters of A3 grade and often poorly removed by
present technologies, this grade of water should, as far
as possible, not be used for potable water preparation.

The WHO International Standards for Drinking Water
indicate a "highest desirable level" and a "maximum

permissible level" for a number of water pollutants. The Commission of the European Communities Directive for Drinking Water distinguishes between a "maximum allowable concentration" (MAC) and "guideline values" for some non-toxic pollutants; exceptions are allowed for the MACs where no danger to health is involved or if the pollutant is of natural origin; the MAC values apply to water at the point of delivery. The interim primary drinking water standards for public water supply in the United States specify only maximum permissible levels for pollutants "at the tap".

A survey of standards for overall parameters is given in Table III.11. It shows that the most universally applied overall parameter for drinking water quality is the turbidity. Up to now no standards have been formulated for the Total Organic Carbon (TOC) concentration; in a report by WHO "Health Aspects of Direct and Indirect Re-use of Waste Water for Human Consumption" a maximum permissible level of 5 mgC/l has been recommended and lower levels up to 1 mgC/l are mentioned for recycled water.

Values for maximum allowable concentrations of inorganic specific pollutants are summarised in Table III.12. The concentration of a specific pollutant theoretically should not change during transport of the water from the treatment station to the tap of the consumer. Considerable increases, however, can occur as a result of dissolution of substances from distribution pipes of galvanised cast iron, copper or lead. Copper and zinc are often allowed at higher levels "at the tap", but lead and cadmium are not because of their high toxicity. Increases may also often occur for certain organic compounds such as organochlorines.

The number of compounds for which standards for specific organic pollutants are enumerated is fairly small and it must be realised that the presence of many compounds is implicitly limited either by the taste or odour restrictions or by standards for other general parameters. The United States 1975 Interim Primary Standards for drinking water include: chlordane 3.0 ug/l MAC, endrin 0.2 µg/l MAC; heptachlor 0.1 µg/l MAC; heptachlor 100 µg/l MAC; toxaphene 5 µg/l MAC; 2,4D 100 µg/l MAC; 2,4,5-TP Silvex 10 µg/l MAC.

Table III.11

STANDARDS FOR NON-SPECIFIC PARAMETERS

Parameter	Unit	(E.E.C. 1975) Surface Water Intended for Drinking Water A3		(E.E.C. 1975) Standards for Drinking Water	(USA 1975) Interim Primary Standards for Drinking Water	(W.H.O. 1971) International Standards for Drinking Water
		Guideline	M.A.C.	M.A.C.	M.A.C.	
pH		5.5-9.0		9.5		6.5-9.2
Chloroform extract	mg/l	0.5			0.7(1)	
Conductivity (20°C)	μS/cm	1000		1250 (**)		
Coloration	mgPt/l	50	200(*)	20 (*)		50
Kjedahl Nitrogen	mgN/l	3		0.5		
Odour Number (25°C)	dil	20		3		Unobjectionable
Oxydizability ($KMnO_4$)	mgO_2/l	30		5(*)		
Taste Number (25°C)	dil			3		Unobjectionable
Total Dissolved Solids	mg/l			1500(*)		1500
Turbidity	Si scale			0.3	1(*)	25

(*) Exceptions allowed

(1) Carbon-Chloroform extract.

Table III.12.

STANDARDS FOR INORGANIC PARAMETERS

Parameter	Unit	(E.E.C. 1975) Surface Water Intended for Drinking Water A3		(E.E.C. 1975) Standards for Drinking Water	(U.S.A. 1975) Interim Primary Standards for Drinking Water	(W.H.O. 1975) Recommended International Standards for Drinking Water
		Guideline	M.A.C.	M.A.C.	M.A.C.	Max. Permanent level
Aluminium	μg/l			50(*)		
Ammonia	mgNH4/l	2	4(*)	0.05		
Antimony	μg/l			10		
Arsenic	μg/l	50	100	50	50	50
Barium	μg/l		1000	100(*)	1000	
Boron	mg/l	1.0				
Cadmium	μg/l	1.0	5	5	10	10
Calcium	mg/l					200
Chloride	mg/l	200		200(*)		600
Chromium	μg/l		50	50	50	
Copper	μg/l	1000	50	100(1)		1500
Cyanide	μg/l		50	50	200	50
Fluoride	mg/l	0.7-1.7		0.7-1.5	1.4-2.4	0.6-1.7
Iron	μg/l	1000		300(*)		1000
Lead	μg/l			50	50	100
Magnesium	mg/l	1000		50		150
Manganese	μg/l	1000		50(*)		500
Mercury	μg/l	0.5	1	1.	2	1
Nickel	μg/l		50(*)	50(*)		
Nitrate	mgNO3/l			50(*)	45	45
Nitrite	mgNO2/l			0.1		
Phosphate	mgP/l	0.3		2		
Potassium	mg/l			12(*)		
Selenium	μg/l	10	10	10	10	10
Silver	μg/l			10	50	
Sodium	mg/l		100(**)	100(**)		
Sulphate	mg/l	150	250(**)	250(**)		400
Zinc	μg/l	1000	5000(1)	100(1)		15000

(*) Exceptions allowed.
{1} After 16 hours detention in mains, M.A.C. for copper 1.5 mg/l and for zinc: 2.0 mg/l.

III.8. CURRENT TREATMENT TECHNOLOGIES
IN LARGE-SCALE USE

The primary objective of potable water managers in most countries has hitherto been the production of clear, pathogen-free waters without observable taste or odour. Thus the emphasis has been on improving methods for the reduction of taste, odour or colour and relatively few studies have been completed in connection with the removal of individual specific and potentially hazardous pollutants. Therefore, the following is only a general analysis of the present position and of the efficiency of existing techniques.

The basis of current treatments is a number of well-established sequences made up of a limited set of basic processes. It should be noted, however, that where there is an increased demand for water, the tendency is to accelerate the rhythm of operations in the existing plant; this increased rate generally has a negative effect on water quality, particularly when the equipment has not been designed for a high throughput rate. In order to achieve this "increased yield" and reduce water quality loss, recourse is made to complementary purification stages, notably adsorption on activated carbon and strong oxidation with chlorine. Solutions to some of these problems have been proposed.

The design of existing plants is generally based on either biological or physico-chemical methods or on combinations of both. Biological plants generally feature a three phase treatment. This begins by "trimming" of the intake waters by passage through bank filters or reservoirs. Purification is completed by various pre-filtering, slow biological filtering techniques, pH control and water conditioning. Where practised, final disinfection has been almost invariably, by chlorination. Figure III.1 illustrates the most frequent processes.

Physico-chemical plants, in contrast, rely on the addition of various coagulating and flocculating agents followed by solids removal and final disinfection. Filtration in plants of this type is generally more rapid and throughput is correspondingly higher. The ever-present danger is, however, that intensive use of chemical reagents is likely to cause directly or indirectly, the release of undesirable chemicals unless the operations of decantation and filtration are perfectly carried out, which may not always be the case.

Supplementary purification may be performed using complementary adsorption and/or oxidation stages at various positions in the treatment chain. Another alternative is the inclusion of a pre-treatment oxidative, high pH stage giving enhanced precipitation of various metallic pollutants.

A. EFFICIENCY OF WATER TREATMENT METHODS
IN REMOVING SPECIFIC POLLUTANTS

As indicated earlier, few studies of removal rates for individual specific pollutants have been completed. Of the available studies, fewer still include comparative data for alternative processes or sequences. Inevitably, therefore, current performance generally has to be evaluated in terms of overall pollution reduction or, at best, in terms of the control of various broad categories of pollutants.

In evaluating removal efficiency, it is also important to recognise the interactive nature of many reagents and pollutants. Thus, current processes, in addition to acting on pollutants can also have a negative effect on water quality. For example, pre-oxidation may be used for improving metallic precipitation but the process is likely to require large quantities of ozone or chlorine if used for waters with high concentrations of organic material, and moreover will inevitably give rise to the formation of significant levels of undesirable by-products like organochlorines which, once formed are very difficult to remove and are likely to reduce adsorption of PAH's and phenols by active carbon.

Removal efficiencies also vary with different parameters of the water to be treated. For example, acid waters are likely to require additional raising of the PH for the removal of various metallic or other pollutants to be efficient.

Illustrative data of removal efficiencies typically achieved by conventional processes are given in Tables III.13 to 16 and also in the matrix in Table III.17. Any table showing removal efficiency of any treatment process can only be a rough guide. The efficiency of treatment varies considerably with:

- the concentration of pollutants in question;
- the form in which pollutants are found, e.g. whether they are complexed or not;
- the nature and concentration of other pollutants present;
- the effect of treatments before and after the considered treatment processes.

143

Table III.13

REMOVAL OF SOME ORGANIC SPECIFIC POLLUTANTS BY SELECTED TREATMENT SEQUENCES

Pollutents	Reduction in %			Reduction in % by advanced methods*
	Biological	Physico-chemical	Mixed	
Whitening agents	20	0	20	55
Complexing agents & chelating agents	20	?	20	20
Cyclic and polycyclic hydrocarbons	50	50	75	100
Emulsified hydrocarbons	25	50	65	95
Ligno-sulphonic acids	25	50	65	85
Nitrogen compounds(1)	20	?	20	85
Phenols	60	60	85	100
Plasticisers	30	60	72	95
Polychlorinated compounds	?	30-60	30-60	90
Surface active agents	80	50	90	100

* Systems assumed to include pre-ozonisation and activated carbon treatment stages in addition to conventional processes.

(1) Ammonia excluded.

144

Table III.14

COMPARISON OF SPECIFIC POLLUTANT LEVELS BEFORE AND AFTER
PASSAGE THROUGH A TREATMENT PLANT, MEASURED IN A
CHLOROFORM EXTRACT (in µg/1)

Treatment plant		Phthalates	Chlor-inated hyd-cbn	Phenols	Chlor-inated Phenols	Phenols steroids	Organo Hg	Nitrosa-mines	Poly-cyclic hyd-cbns	Indolics	Aromatic amines	Total of all spec. poll. present
A	Entry	4.25	2.50	8.75	2.50	1.25	1.00	1.00	7.25	.50	1.25	1090
	Exit	4.25	5.50	1.50	3.50	.25	.00	.00	2.25(1)	.00	.25	580
B	Entry	2.50	3.50	2.00	1.00	.25	.00	.00	3.50	.00	.00	960
	Exit	5.75	5.25	1.25	2.50	.00	.50	.50	1.25(1)	.00	.00	340
C	Entry	2.25	8.50	.75	.25	.00	.00	.00	.75	.00	.25	450
	Exit	4.00	7.50	.75	.75	.00	.25	.00	.50(1)	.00	.75	340

Each figure represents a mean total level, for a given amount of extract, based on four samples.

(1) In other studies, however, polycyclic hydrocarbons have been removed completely during processing.

Table III.15.

AVERAGE % REMOVAL OF TRACE METALS AT
POTABLE WATER TREATMENT PLANTS

Metal	Average % by micro-strainer	Average % by clarifier	Average % by filter	Average % by plant
Analysis by X-ray emission				
Cr	25	35	15	31
Cu	14	26	37	49
Fe	47	51	49	65
Pb	3	27	29	32
Mn	36	30	55	63
Mo	21	8	12	15
Ni	3	40	41	54
Zn	30	36	37	48
Analysis by AA/Wet chemistry				
Cu	1	50	39	55
Fe	15	51	60	81
Mo	2	8	3	11

AA - Atomic adsorption spectrophotometry.

Source: Reference 26.

Table III.16

REMOVAL OF METALLIC SPECIFIC POLLUTANTS BY SELECTED
TREATMENT SEQUENCES

Limit, in µg/l, for drinking water	Reduction in %						Reduction % using high pH treatment
	Biological		Physico-chemical		Mixed		
	pH <6	pH between 7.8 & 8	pH <6	pH between 7.8 & 8	pH <6	pH between 7.8 & 8	
Zn 100	30	75	30	54	30	81	95
As 50	30	92	80	96	80	97	95
Be ?	?	?	?	?	?	?	?
B ?	0	0	0	0	0	0	?
Cd 5	30	50	45	68	45	77	95
Pb 50	30	50	45	99	45	99	99
Ni 50	30	50	30	50	30	64	99
Hg 1	30	50	30	60	30	71	80

(1). Proposed by the Commission of the European Communities in July 1975.

146

Table III.17

Matrix showing degree of removal of specific water pollutants
in waste water and water treatment processes

Degree of removal
Key: H ≥ 80 per cent
M   20-80 per cent
L ≤ 20 per cent

Treatment stage	1			2							3			4		
Process code letter in hazard rating	A	A	B	C	D	E	F	D	E	F				G	H	I
Degree of removal	Settlement	Dissolved air flotation	Coagulation and settlement	Anerobic contact process	High-rate activated sludge	Conventional activated sludge	Extended aeration	High-rate biological filtration	Conventional biological filtration	Two-stage biological filtration	Treatment on grass plots	Sand filtration	Micro-straining	Denitrification	Activated carbon filtration	Chlorination
Bleaching agents	L	L			H	H	H	H	H	H		L	L			
Complexing agents - EDTA	L	L	M		L	L	L	L	L	L	L	L	L	L	L	L
- NTA	L	L	M	M	L	H	H	L	H	H	L	L	L	L	L	L
Cyclic and polycyclic hydrocarbons	M	M	H	L								M	L		H	
Hydrocarbons - emulsified	M	M	H	L	M	H	H	M	H	H		L				
- in storm water	M	M	H	L	M	H	H	M	H	H						
Ligno-sulphonic acids	L	L	M	L	L	L	L	L	L	L		L	L			
Nitrogen compounds - ammonia	L	L	L	L	L	M	H	L	M	H		L	L		L	
- aromatic amines	L	L	L		L	M	H	L	M	H		L			M	
- amides	L	L	L		L	M	H	L	M	H		L			M	
Phenols - monohydric	L	L	L		M	H	H	M	H	H		M	L		H	
- polyhydric	L	L	L		L	M	M	L	M	M		M	L		H	
Plasticizers	L	L	M									L			H	
Polychlorinated compounds - BHC	M					L	L		L	L		L	L		H	
- Dieldrin	M	M			M	M	M	M	M	M		M	L		H	L
- PCB	M			L	M	M	M	M	M	M		L	L		H	
Surfactants - Monyl phenol ethoxylate	L				L	M	H	L	M	H		L	L		H	
- Cationic	L		H	H	H	H	H	H	H	H		L	L		H	
Arsenic	L	L	H	M		M	M		M	M		L			M	
Beryllium												L				
Borate	L	L		L	L	L	L	L	L	L		L	L		L	
Cadmium	M	L	H	M	L	M	M	L	M	M		L				
Lead	M	L	H	M	M	M	M	M	M	M		L				
Nickel	L	L	H	L	L	M	M	L	M	M		L				
Organo-mercury compounds												L	L			

Table columns (as labelled across the top):

- **4.9** — J: Ozonation; K: Coagulation and settlement with — a – lime, b – aluminium salts, c – ferrous sulphate, d – ferric salts, e – activated silica, f – Bentonite clay, g – organic poly-electrolytes
- **4.10** — L: Cation exchange; M: Anion exchange; N: Mixed-bed ion exchange; O: Reverse osmosis; P: Ultra-filtration
- **6** — R: Storage
- **7** — K: Slow sand filtration
- **9** — Coagulation and settlement (see 4)
- **10** — S: Gravity or pressure filtration; H: Activated carbon filtration; T: Activated alumina filtration; U: Bone char filtration; V: Electrodialysis
- **11** — I: Chlorination; W: Chlorine dioxide treatment; J: Ozonation
- **13** — X: Recharge; Y: Distillation

J	a	b	c	d	e	f	g	L	M	N	O	P	R	K	9	S	H	T	U	V	I	W	J	X	Y
L	L		L	L	L	L	L	L				L		H	H	L	H	L	L		L	L			H
L											H	L		L	H	L	M	M	L		L	L			H
L		L						L	L	H	H	H	L	M	H	L	M	M	L		L	L			H
L	H	H										L		H		L	H	M	H		L	L	M		
L	H	M	H	H	H	L	M	M				H	L	H	H	L	M		M		L	L	L		M
L	H	M	H	H	H	L	M	L				L	H	H			M		M		L	L	L		M
M	M	M	M	M								H			L	L	M		L		M	M	M			H
H	L	L	L	L	L	M	L	H	L	H	H	L	M	M		L	L	L	L		H			H	L
								H	L	M	M					L	L							H	M
								H	L	M	M					L	L							H	H
M	L	L	L	L	L	L	L	L	L	M	H		L	H	L	L			M	H	H	M	L
L	L	L	L	L	L	L	L	M	L	M	M		L	H	L	L		M	M	H	M	M
L	L	L	L	L	L	M	L	H	L	L	M			H		M		L	L	M	M	M
L	L					M		M	M	L	M	L	H	H	M	L	L	M	H	M
M	M	M				M		H		L	M	L	H	H	M	L	L	M	H	M
L								H	M	L	M	L	H	H	M	L	L			M
L	H	H						H	L	M	M	L	H	M	M	L	L			M
L								H	L	H	H	L	H	M	M	L	L			H
L	H	M	M	H	H		L	H	H	H	L	L	L	L	M	M	L	L	L	M	H
L								L	L	L	L				L	L	L
L	L	L	L	L	L	L	L	L	H	H	M	L	L	L	L	L	L	L	L	L	L	L	M
L	M	H	L	H	M	L		L	H	L	H	M	L	L	L	M	M	M	L	L	H	H
L	H	H	H	H	H	L	H	H	L	H	H	L	M	L	L	M	M	M	L	L	H	H
L	H	H	L	H	L		H	L	H	H	L	L	L	L	M	M	M	L	L	M	H
L	H							L	L	L	L	H	M	L	L	L	L

Because of the interaction within treatment processes, it is better and more appropriate to consider the effectiveness of a treatment chain as a whole.

Removal efficiencies for elemental contaminants under a variety of regimes are shown in Tables III.15 and 16. Depending on the pollutants in question and the operating conditions the removal rates may vary between 0 and 99 per cent (Boron and Lead respectively). In general, removal efficiencies between 50 per cent and 80 per cent are typical for most elemental contaminants. Efficiency may, however, vary widely with differing conditions as can be seen in Table III.16 which gives an indication of the higher removal efficiency obtained at high pH values.

In evaluating these and other removal efficiencies, it is important to be aware that most pollutants are present in a variety of forms. Thus a pollutant may be firmly fixed in its original mineral matrix, surface adsorbed, adsorbed on colloids, organically combined or present in a range of inorganic combinations. The removal efficiencies (and health implications) for differing categories are unlikely to be the same.

Removal efficiencies for some groups of organic compounds under three treatment regimes are given in Table III.13. Reported efficiencies range between 0 and 100 per cent but most are between 20 and 60 per cent. Values for a further 10 groups of compounds are given in Table III.14. It will be noted from this table that levels of chlorinated phenols, chlorinated hydrocarbons and phtalates all actually increase during the purification process. In every case, however, the range of efficiencies noted is wide with phtalates for example, increasing during processing at two treatment plants, remaining the same at a third. Similar variations have been noted in other studies and it is evident that the area is one requiring a major investigative effort. Table III.13 indicates the improvements which can be brought about by the application of additional pre-ozonisation and high pH stages. On the other hand the application of strong oxidants before the final stage of treatment is now considered inadvisable as already discussed.

One particularly important point, in considering removal rates for organic pollutants, concerns the various refractory compounds resistant to current purification processes. These materials may be refractory because of high biological stability (e.g. PCBs), their toxicity to micro-organisms, and/or because they are poorly removed by the various physico-chemical processes adopted. An especially unwelcome aspect of the refractory problem concerns the ever-increasing proportion of these substances in many intake waters. Thus, whilst

149

biological treatment of wastes for example is becoming widespread, it is evident that such methods cannot reduce levels of bio-refractories. The materials will consequently tend to increase relative to other organics. An estimate has already been mentioned in connection with the Rhine where the total organic carbon (TOC) is expected to be composed of a high proportion of bio-refractory products in the near future(9). This demonstrates that the role of BOD must be drastically reconsidered. Nevertheless, even if TOC is more reliable as a measure, BOD still gives useful indications, especially when associated with TOC for the assessment of bio-refractory organics. Where refractories are present, treatment methods such as carbon adsorption are desirable.

III.9. POSSIBLE IMPROVEMENTS
IN TREATMENT PROCESSES

Without sufficient data on the different types of
raw waters and the efficiency of existing techniques,
it is not possible to establish "in principle" the precise number and type of supplementary processes needed
to achieve given purity levels. Provision of additional
processes is likely, however, to be of value because of
greater efficiency and safety over single-barrier
systems. The likely efficiency of some typical multi-
barrier systems used for potable water is summarised in
Tables III.13 and Figure III.2 which show that over
90 per cent efficiency can be achieved for certain pollutants. However efficiency for many other pollutants may
remain low (chelating agents, are only removed at
20 per cent). Increasing the number of barriers in such
systems is also likely to imply some increase in both
capital and running costs.

Treatment costs for potable water supply are a relatively modest part of total costs, and even a significant
improvement in the treatment process would increase the
price of water by only a few per cent. Considering the
small contribution of the cost of water in their overall
budget, it seems likely that consumers would often be
prepared to pay a little more for much improved water
quality.

The following paragraphs briefly list and discuss
options based on alternative uses of chemicals, new processes and different management approaches. The options
are grouped into short, medium and long time scale
improvements.

Short-term improvements

a) <u>Management options</u>: It might often be possible
to obtain better results using more carefully and at an
optimum rate, the <u>classical processes</u> already in operation. An essential measure, in particular, is rigorous
clarification of the water with a maximum reduction of
suspended solids which can be achieved, for instance, by
better settlement and slow filtration, and can by itself
provide considerable reduction in organic and inorganic
materials. Sometimes, however, a noticeable deterioration in water quality has occurred because of a desire

to increase the "yield" (throughput) of the plant by accelerating process flow rates at the expense of treatment time. This has generally decreased the effectiveness of pollutant removal.

b) <u>Alternative use of chemicals</u>: It is a vast field where careful experimentation is needed in order to determine: the efficiency and reliability of the chemical in full-scale operation; its acceptable cost; and, the absence of disadvantages from health or nuisance effects. The latter requirement should be particularly stressed.

Modern technology has enabled a great number of new adsorbents, and related coagulants based on minerals such as silicates or on polyelectrolytes to be developed. Many of these adsorbents can be made chemically specific, such as the scavenging resins or, they can be specific through biological effects.

A decisive stage in eliminating specific pollutants being the removal of suspended matter, the chemical reagents used as either coagulants or flocculants in this step play a critical role. Fundamental knowledge of these processes has greatly increased, and progress in chemistry has resulted in new and more efficient products but for some of these (such as synthetic organic chemical polymers) there are doubts about their biological properties. Other newer coagulants, which appear more appropriate, include polyhalogenated aluminium compounds and other mineral complexes. Careful choice of appropriate coagulant/flocculant is essential.

c) <u>Alternative processes</u>: Provision of additional stages of treatment by activated carbon, either in beds or columns, can often be arranged. This material has an extremely high surface area to volume ratio giving a powerful adsorptive capacity and resulting in the rapid removal of various pollutants especially non-polar ones such as aromatics and saturated aliphatics. Inorganic substances by contrast are poorly removed. If used after chlorination, the carbon beds may be quickly saturated by organochlorines and their efficiency reduced. Carbon adsorption should be used at the final stage of treatment to eliminate residual organic compounds just before disinfection (with a light application of chlorine dioxide, ozone or chlorine). The systems known as "biological filtration", on carbon in particular, but also on sand or soil, is a well proven and very effective technique which should be developed and generalised.

Medium-term improvements

The alternative measures listed here may overlap with either the short or longer-term options, and will require extensive reorganisation of the plant and improved operator skills.

a) <u>Management options</u>: These will generally depend on improvement of supply sources and better turbidity control in treatment plants. They include:

   i) establishment and implementation of a hierarchy of water use, reserving and allocating the high quality waters for potable use, the low quality raw waters being gradually abandoned and allocated to other less demanding uses.

  ii) maintaining an improved protection of waters used for drinking, including strict control of possible pollution sources.

 iii) optimisation of potable water treatment plant operation

  iv) re-location of water abstraction points. This is a general problem in Member countries as many abstraction points for potable water are situated in urban and suburban areas because towns have spread outwards in recent years. A necessary measure - frequently discussed - is to move the abstraction points upstream of the urbanised zones. For this to be fully effective, several conditions must be satisfied:

- the displacement should be sufficient so that it need not be repeated within a few years;
- the catchment upstream of the new abstraction site must be protected against polluted discharges (industry, municipalities);
- raw water transported must be sufficiently pure. Transporting untreated water over long distances may sometimes lead to side effects such as deposits of mud and proliferation of organisms. Up to now, chlorination has been used as a method of control, but this should be discouraged because of the early formation of organochlorines which cannot be easily removed later. A rational solution is to have sufficient clarification of the raw water at the upstream abstraction point together with a final treatment at the water plant, which can be much simpler.

b) <u>Use of chemicals</u>: For example, greater use of ozone is possible. Optimum application should be carefully controlled in order not to interfere with other operations. A light application should be used at the end of treatment and could be usefully associated with filtration on activated granulated carbon. Finally, the light application of a "remanent" "bacteriostatic agent" (chlorine dioxide, chloramine chlorine) may be desirable to protect the network.

c) <u>Alternative processes</u>: Recent advances have significantly extended the scope for ion exchange techniques
and include development of new (synthetic or natural)
materials, easily regenerated and highly selective for
obnoxious ions such as ammonia, and nitrates. Ion exchange also seems promising for removing organics and may
become complementary to activated carbon adsorption(12).
The comparatively high cost of these techniques may be a
limiting factor in their development. Progress in micro-
biology will permit significant improvement in biological
treatment, which is quite efficient not only for nitrification and/or denitrification but also for removing many
organic substances. Biological filtration is particularly
interesting because it combines physico-chemical and biological processes. Filtration on granular activated carbon is the best example of such a treatment and considerable improvements may be expected from optimising this
phenomenon. Natural soil filtration is used successfully
in some countries and, in view of its qualities, it should
be given more attention. Storage in reservoirs is also a
useful technique and enables purification by physico-
chemico/biological means; this procedure is not very widespread and it should be encouraged. Considerable progress
in potable water production will probably come from a
better sequencing of conventional treatments.

<u>Longer-term alternatives</u>

In the longer term (i.e. 15 years) it is not likely
that water supply and potable water problems in OECD
countries will be resolved by "advanced" new technologies.
It is more realistic to consider that water problems will
still have to be resolved by careful management mainly
using (improved) classical methods and technologies.
This is not intended to discourage or minimise the support
that new advanced technologies may provide, but only to
limit a misleading tendency to overestimate them. In
fact, a number of new approaches, which 10 years ago
seemed very promising (i.e. sea water desalination) now
appear, because of the uncertain energy outlook, to have
limited prospects in OECD Member countries, as a result of
their energy intensiveness. Possible new developments
include:

a) <u>Chemicals</u>: Scope for significant improvements is
considerable in the field of both pollutant removal and
disinfection.

b) <u>Processes</u>: Electrodialysis and reverse osmosis
are beyond the experimental stage and already in full-
scale use. Nevertheless, their cost is high and they
cannot compete where classical methods can be applied.
Other membrane technologies, such as ultra-filtration, are
also being developed and significant improvements may be
expected. It is likely that it is the "energy

intensiveness" of these technologies, which will be the deciding factor in their wider adoption.

c) <u>Management options</u>: Considerable evolution should take place in this field; the main management options are in three essential directions, namely <u>better raw waters, better treatment, and more rational distribution</u>.

i) <u>Selective use of better quality raw waters</u>. This is probably the most important and effective measure. In the past, high quality waters were traditionally reserved for potable water and the increasing tendency to utilise nearby low quality waters for potable supplies is often a relatively recent problem. Several factors have contributed to this situation - the actual deterioration of a number of resources; increased demand; over-estimation of technical possibilities; and a tendency towards short-term investment.

The prime importance of using high quality raw water has been emphasized by the experts and the following approaches will be the most common:

- More rigorous pollution control in water bodies used, or to be used, for potable water. Nevertheless, in many densely urbanised and industrialised river basins, the sources of contamination are so extensive that it is unlikely that the water quality will recover, within an acceptable time to a desired level.
- Allocation of the best quality sources for potable use. The problem of "water rights" and, for instance, frequent pre-emption by industry of large volumes of high quality waters may need the enforcement of stricter legislative, regulatory and economic measures.
- Where it is still possible the development of new good quality resources is, of course, an appropriate solution. Unless these resources can be found within a relatively short distance, long distance transportation of high quality waters is still the safest policy. This method (which has been practised since antiquity) is, in fact, successfully used in a number of Member countries. The investment, once made, can be amortized over a long time, and the much-reduced cost of treatment compensates for that of transportation. The social benefit of supplying high quality water is of course a determining factor.

ii) <u>Management of treatment</u>: Many improvements can be envisaged for the management of potable water treatment plants, and new technologies complementary to the treatment processes would be particularly valuabe. An example is the use of automatic

155

measurement and monitoring devices to indicate qualitatively and quantitatively the presence of different classes of pollutants. These devices should monitor the raw water intake, the potable water before distribution, and the water at key points in the treatment, where possible. The computerised information should then permit instant adjustment of operations in relation to the presence and removal of pollutants. Sensitive and quick tests of overall toxicity would also be useful before distribution. (For instance, the use of a cytoxicity test is described in Annex III.2).

iii) <u>Rational distribution of potable water in relation to availability of good quality raw waters</u>:

a) Distribution strategies can be flexible when there is sufficient high quality raw waters for all users, but when such supplies are limited, these should primarily be reserved for potable water (domestic water supply) and other high quality requirements (food processing).

b) In practice, however, in order to meet demand, increasing quantities of low quality raw waters generally have to be abstracted, at the expense of potable water quality.

c) Two different positions exist with regard to this dual network (i.e. domestic, industrial);

- on one hand there is concern in some countries that the existence of two networks creates a risk of confusion in connecting water pipes, even if this risk is extremely low;
- on the other hand the existence of a single network may cause a permanent reduction in drinking water quality and consequently a permanent health risk.

d) Finally, the exact profile of potable water distribution should be based on the degree of scarcity* of good raw water resources and should

---

* Where water supplies are particularly difficult (in arid regions, islands and over-populated areas) and where good quality waters are insufficient to supply all domestic needs, it may be necessary to keep these scarce, good resources for demanding domestic needs (potable water/kitchen/bathroom). "Multi-purpose" water (duly disinfected) can be supplied for all other less demanding uses (toilets, laundries, garages, gardens, urban, industrial etc.) and recycled water can be used for this purpose. Various alternatives of such systems can be found in several places (e.g. Japan, Hong Kong, Bermuda, and

(Continued on next page)

take into account the balance of the various regional and local needs. More rational and longer-
term planning of potable water supply and distribution would not only better fulfil health and social
requirements but would, in most instances, be less
costly for the society as a whole.

(*. Continued from previous page).
various industrial estates in the United Kingdom). Contrary to certain criticisms, such systems, if well
managed, do not pose health risks since even non-
drinking water must be disinfected.

III.10 CONCLUSIONS

A detailed survey should be made of water resources in Member countries in relation to the most relevant parameters, including specific pollutant contents. This should be done with a view to resource reallocation locally, regionally and nationally, according to a strict hierarchy, with the highest quality reserved for human consumption.

The increasing use of low quality raw waters for potable water should be discouraged, as in practice the water prepared is less satisfactory in terms of human health and taste. A number of potentially hazardous micropollutants in these waters may pass unremoved through the treatment plant, while relatively harmless pollutants may be transformed into toxic chemicals (such as organochlorines) by the reagents used in the treatment itself. Consequently, all possible efforts should be made to uptake the supply of good quality raw waters locally, or if necessary, from longer distances; aquifers should also be brought more actively into the management system, provided nitrate contamination can be prevented.

When good quality raw waters are in short supply, the tendency to supply all users (even those of industry which do not require such quality) from the potable water networks should be reconsidered as it leads almost invariably to increased uptake from lower quality sources in order to meet total demand; this inevitably results in decreased potable water quality. The policies for potable water supply and distribution in general, should be established primarily in the light of health and social requirements and not only from the restricted engineering and economic viewpoints.

Available regulatory, managerial and technical controls should, where necessary, be more thoroughly enforced in order better to safeguard surface and underground waters used for drinking water supply. Catchment areas must be carefully protected. Moreover, aquifers should be recharged only with adequately purified water.

All water resources used or likely to be used for water supply should be regularly monitored. The frequency and comprehensiveness of monitoring should be adjusted to the hydrological circumstances (flow velocities, turnover time of the water), to exposure to specific pollutants (regular and accidental), and to the type and intensity of use. The parameters monitored should reflect the anticipated specific pollutants.

Biological and chemical monitoring of water quality, at potable water treatment plants, should be carried out continually on influent raw water, and on the water prior to distribution to the public. Points of consumption should also be monitored to control contamination derived from the distribution system.

During treatment, particular attention must be paid to the possible formation of undesirable by-products through the use of impure chemicals, excess quantities of reagents or an unwise sequence of unit processes. For instance, in order to limit the formation of by-products (such as organochlorines, resulting from the use of chlorine) oxidents in general should be used more carefully and limited to final disinfection when organic precursors have been minimised.

At present, potable water standards only cover a limited range of mainly inorganic specific pollutants. However, as many organic pollutants are particularly hazardous, it follows that water meeting "classical" standards will not necessarily be safe. Member countries should urgently develop and adopt a range of new individual and comprehensive standards covering:

- raw water,
- water at the point of consumption,
- reagents used in preparation,
- materials used in construction of plant, equipment and distribution systems.

Water softening in the home should be more strictly controlled and restricted to such equipment as water heaters and washing machines. Softened water should not be used for regular human consumption because of its possible health impacts, i.e. noxious change in ion balance, by increasing undesirable sodium and removing useful magnesium and calcium. Furthermore over-softening may lead to the dissolution of hazardous metals (cadmium, lead) from pipes.

Water distributed from the treatment plant should be non-aggressive in order to minimise corrosion and dissolution of hazardous materials, such as metals and asbestos from the distribution network. For health reasons also, neutralisation should not be carried out with caustic soda but with lime.

Potable water plants should be equipped with sufficient storage basin capacity. This provides increased flexibility in plant operation as it can fulfil a number of useful purposes for both raw and finished waters, thus permitting a higher and more constant quality. Moreover,

the design of the intake system should be carefully
undertaken to avoid regular or accidental pollution,
especially with surface waters.

The desire to increase the throughput of water
treatment plants may frequently result in raising the
rate of unit processes; but, unless the processes have
been designed and optimised for this rate, the result is
lower final effectiveness in the removal of specific
pollutants. It is necessary to reverse this trend, and
to revert to rates of maximum effectiveness for pollutant
removal.

CLASSIFICATION OF RIVERS ACCORDING TO THE DEGREE OF POTENTIAL ARTIFICIAL POLLUTION*

A classification of rivers according to the degree of potential artificial pollution and probably also in relation to the necessary treatment for production of drinking water can be set up according to a Potential Pollution Index (PPI) which is defined as follows:

$$P.P.I. = \frac{N \times G.N.P./cap.}{Q \times 10^6}$$

where:
P.P.I.   = Potential Pollution Index for a specific river site and year of observation
N   = Number of people living in the drainage area considered
G.N.P./cap.   = Average Gross National Product/ Capita (U.S.$) for the population of drainage area considered
Q   = Yearly average river discharge $(m^3/sec.)$

A study by the W.H.O. International Reference Centre for Community Water Supply provides water quality data on roughly one hundred river basins. According to these data a natural relationship between the P.P.I. and the ratio of the drainage area of the river and its discharge can be derived for countries with limited industrialisation having a G.N.P./cap. below U.S.$1,000. This relationship is tentatively reported to be as follows:

$$\log (P.P.I._n) = 1.75 \log \left\{\frac{D.A.}{Q}\right\} - 3.75$$

where:
$P.P.I._n$ = Potential Pollution Index due to natural causes of pollution
D.A.   = Drainage Area of river in km^2
Q   = Yearly average river discharge $m^3/sec.$

Based on this "natural" factor and the actual P.P.I., rivers can be classified as:

---

* Zoetman, 1973 (8).

Figure III.1

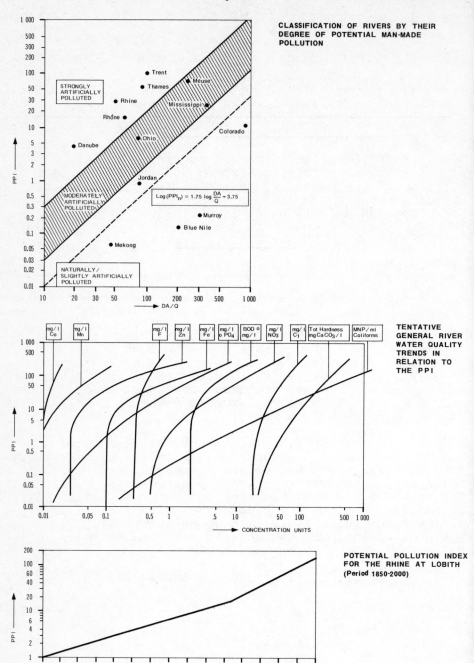

CLASSIFICATION OF RIVERS BY THEIR
DEGREE OF POTENTIAL MAN-MADE
POLLUTION

$$\text{Log}(PPI_n) = 1.75 \log \frac{DA}{Q} - 3.75$$

TENTATIVE
GENERAL RIVER
WATER QUALITY
TRENDS IN
RELATION TO
THE PPI

POTENTIAL POLLUTION INDEX
FOR THE RHINE AT LOBITH
(Period 1850-2000)

162

Naturally/slightly artificial polluted -
P.P.I. 3 x $P.P.I._n$;
Moderately artificially polluted -

P.P.I. between 3 x $P.P.I._n$ and 30 x $P.P.I._n$

Strongly artificially polluted - P.P.I. 30 x $P.P.I._n$

The result of this type of classification is illustrated in the following figure.

Annex III.2.

NEW METHODS FOR DIRECT DETERMINATION
OF CYTOTOXICITY OF WATER

Techniques exist for direct determination of cytotoxicity of water, without previous extraction or concentration of contaminents. The extraction or concentration
techniques (i.e. chloroform extraction) are often responsible for the introduction of new micropollutants.
Some extraction techniques can also be selective in their
action.

The principle of these methods is to determine,
in-vitro or in-vivo, the action of toxic water-contained
substances on cellular syntheses. Their main advantages
are:

- high sensitivity because cellular synthesis can
  be significantly reduced by only traces of toxic
  substances without cellular death. Under these
  conditions lethal doses techniques, using large
  quantities of toxics and a long incubation time
  (3 to 7 days), are no longer required and are
  replaced by the 50 per cent Inhibitory Concentration (IC_{50}) concept;
- very fast response (for a biological method)
  varying between a few minutes and 24 hours;
- an important range of syntheses enabling a wide
  toxicological survey which could only otherwise
  be obtained from a complete study with several
  animal species;
- simplicity of operation - it requires in-vitro
  manipulation of only a few millilitres.

These methods are divided into three groups:
methods involving suspended human cell culture; sub-
cellular fractions; isolated cellular and viral enzymes.

I. METHODS INVOLVING SUSPENDED HUMAN CELL CULTURE

In such studies suspended Hela S_3 cells have been
used. The action of various concentrations of toxic
substances on cellular RNA, DNA and protein syntheses
was first determined. In the same way their action on
cellular permeability and oxygen consumption has been

studied. The 50 per cent Inhibitory Concentration was
determined in each case for no more than 120 minutes
(1, 5).

A direct cytotoxicity study on river water (2, 5)
and treated waters* can be undertaken by quantitative
DNA or RNA synthesis determination and this has recently
led to a new micromethod(2).

II. METHODS INVOLVING SUB-CELLULAR FRACTIONS

The action of several toxic substances on ATPase
and Cytochrome-oxydase activity in purified mitochon-
drias was determined as well as their action on isolated
nuclear DNA synthesis and on microsomal ATPase.(1).
Although some of these methods showed a great sensitivity, technological problems prevented them being used
for direct determination of cytotoxicity of water.

III. METHODS INVOLVING ISOLATED CELLULAR
AND VIRAL ENZYMES

The in-vitro action of several toxic substances on
cellular isolated enzymes or on structural viral enzymes
has been studied. Among the studied enzymes /ATPase,
Thymidine-kinase, Uridinekinase, Alkalin Phosphates,
DNA-polymerase, RNA-polymerase (1, 5)7 Alkalin phospha-
tase allows the detection of cyanide in treated water
in concentrations as low as 8 µg/l(5). This very simple
technique is almost 10 times more sensitive than the best
chemical techniques for cyanide detection.

Because of interference phenomena, (currently being
investigated) difficulties still remain in the use of
these techniques for direct determination of river water
cytotoxicity".

Among the proposed techniques, those involving DNA
or RNA cellular synthesis have already been used for
direct and quantitative cytotoxicologic control of river
water (2, 5). The other techniques are also of interest
in view of the need for fast and quantitative determination of effluent cytotoxicity.

---

* Unpublished results (i.e. personal communication).

REFERENCES FOR ANNEX III.2.

1) BAZIN, M.; I. CAZENAVE; C. DANGLOT; R. VILAGINES.
Etude de la Cytotoxicité des eaux. II. Etude de
l'action de différentes substances toxiques sur
plusiers synthèses ou enzymes cellulaires et
virales. JOURNAL FRANCAIS D'HYDROLOGIE, (In Press)

2) CAZENAVE, I.; DANGLOT, C.; VILAGINES, R.
Détermination quantitative de la cytotoxicité des
eaux. Submitted for publication. C.R. Acad. Sci.
Paris.

3) DANGLOT, C.; VILAGINES, R.
Simple methods for direct determination of water
cytotoxicity. (In Press)

4) FERRAN, R.; MAZZA, M.; PAYEN, P.
Utilisation des techniques de multi-détection pour
l'analyse des polluants organiques des eaux.

. Communication - Association Pharmaceutique
Française pour l'Hydrologie - 21 octobre 1977.

. Journal Français d'Hydrologie, Volume 7(22)
(In Press)

5) PLICHON, D.; CAZENAVE, I.; DANGLOT, C.; VILAGINES, R.
Etude de la cytotoxicité des eaux. I. Mise au point
de methodes autorisant la mise en évidence de la
cytotoxicité des eaux sans concentration préalable.

. JOURNAL FRANCAIS D'HYDROLOGIE, Volume 7(19)
pp. 19-32.

N.B.: The JOURNAL FRANCAIS D'HYDROLOGIE is edited
by the ASSOCIATION PHARMACEUTIQUE FRANCAISE
POUR L'HYDROLOGIE, Faculté de Pharmacie de
Paris V, 4, Avenue de l'Observatoire, 75006,
Paris (France).

REFERENCES

1. Harris R.H., Paage T. and Reiches N.A.   Pers. Comm.

2. World Metals Bureau Bulletins.

3. National Academy of Sciences.   Drinking Water and Health, 1977.

4. Science, Vol. 196, May 6, 1977.

5. Burnham, A.K. et al.   Anal. Chem. 44 1 139, 1972.

6. Shapely D., Science 195 42 79.   P. 658. 1977.

7. Penry H.M. et al.   Proceedings Society for Exp. Medicine, Vol. 136, P. 1240.   1971.

8. Zoeteman, B.C.J.   WHO.   Tech. Paper Series b. 1973.

9. Stumm, W. and Roberts, P.V.   3 Arbeitstagung Düsseldorf des Int. Arbeitsgemeinschaft der Wasserwerke im Rheineinzungsgebeit 1973 (Condensatorweg 54, Amsterdam - Slotesdisk, Netherlands).

10. WHO Reference Centre for Community Water Supply. Technical Paper 7.

11. Environmental Science and Technology, 11th April, 1977.

12. Packham, R.F. (WRC, Medmenham, U.K.).   Personal communication.

13. Goldberg.   Lancet April 2.   1977.

14. Anon.   Environmental Science and Technology, 11th January, 1976.

15. Wade, N., Science, Vol. 196, P. 1421.   1977.

16. McConnel, G.   I.C.I. Ambergate Scheme.   April 1976.

17. Page, A.L.   Conference on Heavy Metals in the Environment, Toronto 1975.

18. Reuse of Vehicle Tyres.   HMSO, London.   1976.

19. Zoeteman, B.C.J., Sensory Assessment and Chemical Composition of Drinking Water.

20. Masschelein, W. Cebedeau/Becewa. March 1977.

21. WHO. Reference Centre for Community Water Supply. Technical Paper 9.

22. Lead in Drinking Water. Pollution Paper 12, HMSO, 1977.

23. Cadmium and the Environment, OECD, Paris 1975.

24. Dressman, R.C. and McFarren, E.F. Vinyl Chloride Migration - PVC Pipe. EPA, Cincinnati. 1976.

25. R.F. Packam, U.K. Water Research Centre. Personal communication.

26. Zemansky, G.M. Journal AWA, Vol. 66, P. 606. 1974.

27. Protection of the Environment by Control of Biphenyls, OECD (76) 86. 1976.

28. Cunningham, H.M. & Pontefract, R.D. Asbestos Fibres in Beverages and Drinking Water. Nature 232, 332, 1971.

29. Hallenbeck, W. and Hesse, E. Overview of mechanisms by which asbestos could enter drinking water. Presented at SOEH Conference on Occupational Exposures to Fibrous and Particulate Dust and their Extension into the Environment. December 4-7, 1977.

30. Wigle, D.T. Cancer Mortality in Relation to Asbestos in Municipal Water Supplies. Arch. Environ. Health 32, 185. 1977.

31. Levy, B.S., Sigurdson, E., Mandel, J., Laudon, E. and Pearson, J. Investigating possible Effluents of Asbestos in City Water: Surveillance of Gastrointestinal Cancer in Duluth, Minnesota. Am. J. Epidemiol. 103, 362, 1976.

Part IV

CONTROL OF ORGANOCHLORINATED COMPOUNDS IN
DRINKING WATER

(A Practical Example of a Control Strategy)

IV.1. INTRODUCTION AND SUMMARY

A. OBJECTIVES OF THE STUDY

The relatively recent discovery of organochlorinated compounds in drinking water has raised considerable concern in a number of Member countries because of their possible health effects. Although these effects have not yet been clearly assessed, there might, nevertheless, be a problem arising from long-term exposure of the consumer to these products. Organochlorines present as pollutants in raw waters are, in general, persistent and rather poorly eliminated by potable water treatment processes. Moreover, the chlorination treatment itself is generally the most important source of these compounds in potable water supplied to the public.

The Environment Committee of the OECD has underlined the urgency of the issue and requested an assessment of the most appropriate measures to resolve the potential problems in this field. The following mandate was given:

1. Review and compare the occurrence, levels, and origins of organochlorines in raw, treated and tap waters (from underground and surface supplies). Evaluate existing knowledge regarding the formation of organochlorines in raw, treated and tap waters (from underground and surface supplies). Evaluate existing knowledge regarding the formation of organochlorines by chlorination processes in current use.
2. Review existing knowledge on potential hazards of halogenated organics in drinking water.
3. Discuss possible alternatives for chlorination treatment. Select and evaluate appropriate types and sequences of treatment processes which do not lead to the formation of undesirable by-products in potable water.
4. Evaluate the comparative cost of various alternatives and possible impacts on investment, operation costs and final water pricing.
5. Discuss the comparative advantages and disadvantages of the alternatives envisaged, and propose appropriate measures regarding the different conditions which may be encountered in Member countries.

B. NATURE OF THE PROBLEM

Drinking water production can be divided into three stages: raw water abstraction, treatment processes and distribution processes. Significant contamination can occur at any of these stages where treatment chemicals are applied. The quantity and variety of synthetic organic chemicals formed during the drinking water treatment process can exceed appreciably the amounts arising from raw water contamination.

The primary concern in traditional drinking water treatment has been to control micro-organisms which cause water-borne diseases (such as typhoid and cholera) and to provide an aesthetically acceptable water (taste, odour, colour). This goal has largely been achieved by the use of chlorine and other oxidants in conjunction with other treatment processes. Recently, however, there has been increasing concern about the presence of chemical pollutants in drinking water and their possible health hazard.

With new analytical techniques and instrumentation, such as gas chromatography and mass spectrometry, several hundred specific organic pollutants have been identified in low concentrations(*) in various drinking water supplies. These compounds originate from sources such as industrial and municipal discharges, urban and rural run-off and the drinking water treatment itself. Concentrations of these pollutants vary from virtually nil in drinking water drawn from protected groundwater to substantial amounts in drinking water derived from contaminated surface waters and contaminated groundwaters.

Organic substances present in the raw waters for potable water preparation are of a synthetic and non-synthetic (including natural) origin. A large variety of synthetic organic chemicals may contaminate waters, especially surface waters in industrialised areas. They generally range from low molecular weight hydrocarbons, such as monocyclic aromatic compounds (benzene, toluene) to high molecular weight polycyclic aromatic or polymeric compounds. They may be simple halogenated compounds (such as carbon tetrachloride and trichloroethylene) or complex ones (such as many pesticides and PCBs).

The non-synthetic organic substances consist mainly of humic and fulvic acids and originate from the decomposition of vegetable material as well as from domestic, agricultural and industrial pollution; they are generally the largest fraction of the organics present. Organic

_____

* Typically from 0.01 - 100 microgram/litre (µg/l).

substances in general play a key role as "precursors"(*) because they may interact with oxidants to form by-products.

The parameters currently used in most Member countries for controlling drinking water treatment plants do not generally include measurement of contamination due to organics and organohalogen by-products, but a number of countries envisage doing so. There is a clear need for suitable test procedures and also for operational control of the processes to minimise the formation of these by-products.

So far, four analytical techniques are available to measure actual or potential organochlorine contamination: 1) Total Organic Chlorine (TOCl), 2) trihalomethane (THM) analysis, 3) Total Organic Carbon (TOC), 4) oxidant demand measurement. Techniques 3) and 4) give no indication of the by-products that may be formed during treatment with oxidants but are a measure of the precursors present in the water. Trihalomethane analysis is becoming widely known and is within the analytical capability of at least major water treatment works' laboratories. However, it measures only a small part of the total halogenated by-products formed, but can be considered as a "marker". TOCl is the most comprehensive and relevant test and should be developed as a standard test (it does not indicate, of course, which individual chlorinated compounds are present).

Potable water treatment may considerably increase the content of synthetic chemicals in drinking water and recent studies in many countries indicate that the large number of halogenated products formed by chlorination are often a major portion of the identifiable synthetic chemicals in drinking water. During purification, additional contaminants are therefore being formed and these by-products are found especially in drinking water derived from water containing precursors when treated with chlorine; they can be present at concentrations of up to several hundred microgrammes per litre.

The organohalogens(**) formed are mainly organochlorinated compounds but brominated and iodinated compounds can also be present. Only a portion (about 20 per cent) of the organohalogens present in drinking water can currently be identified(***). These are mainly the volatiles such as the trihalomethanes or THMs, which include chloroform. The other identified compounds which

---

* See Annex IV.1.
** See Annex IV.1.
*** See Figure A (Part IV).

may originate from the raw water account for about 2 per cent but represent a large number of compounds (chlorophenols, PCBs, pesticides, etc.).

The non-volatile compounds (up to about 80 per cent) are difficult to identify with the current analytical techniques (gas chromatography, mass spectroscopy). They represent a large number of compounds and some may be of greater toxicological significance than the identified portion (THM). Their overall level in water can be measured by the TOCl test (Total Organic Chlorine). The proportion of the non-volatiles to volatiles can vary widely.

The total amount of organohalogens reaching the consumer may be higher than the amount measured in the water leaving the treatment plant, because these chemicals continue to form in the distribution system as long as precursors and chlorine are present. By-products may also be formed when using alternative oxidants such as ozone or chlorine dioxide but very limited knowledge exists on these up to now.

At the low individual concentrations at which some organic compounds may occur in drinking water, the primary concern is for a potential contribution to chronic health risks, e.g. cancer. Although the specific causes of cancer are not yet fully understood, there is growing agreement among scientists that exposure to carcinogenic contaminants in man's total environment which includes occupational activities, life style (especially smoking), food, water and air, may contribute to the incidence of cancer which accounts for up to one-third of annual mortality in OECD countries. Thus, several public health agencies have adopted policies limiting human exposure to carcinogens to the maximum extent feasible.

Many organohalogenated compounds may be found in drinking water at low concentrations. Even at the concentration of some microgrammes per litre, the aggregate exposure to such chemicals from a lifetime of water consumption contributes a potential risk to human health by adding to total exposure to hazardous chemicals. In addition, not only the unique exposure to each of these compounds separately should be of concern, but also the added possibility of synergistic effects enhancing associated risk should be considered. Furthermore, certain segments of the population are at greater risk because of age, physical state, environmental stresses and possibly genetic disposition.

The assessment of the effects of synthetic organic chemicals on man is based on animal tests which necessarily use higher levels of exposure than those normally encountered in the environment and on epidemiological studies using statistical data on human diseases and mortality. In 1976, the U.S. National Cancer Institute published a study which showed that under laboratory conditions, cancer was caused in rats and mice by daily exposure to high doses of chloroform.

The actual effect on humans from drinking water containing chloroform at low levels over a long period of time is unknown at the present time. Long-term toxicity tests carried out in France (on mice and rats), using concentrated chloroform extracts of micropollutants from chlorinated drinking water, showed a significant increase in the incidence of various types of malignant tumours. Similar tests carried out with extracts from non-chlorinated groundwaters showed no adverse effects.

Various epidemiological studies have been made to explore the association between organohalogens, or some surrogate parameter found in drinking water, and various types of cancer. Epidemiological investigations in the United States have indicated correlations between increased cancer rates and areas where poor quality surface waters receiving chlorine treatment are the source of drinking water supply. An epidemiological study carried out in the Netherlands, on 4.6 million inhabitants, has indicated that in areas where drinking water is prepared from surface waters of poor quality, which are chlorinated, a higher cancer mortality rate was found (especially esophagus and stomach) than in areas where it is prepared from groundwaters of good quality and generally not chlorinated.

Epidemiological studies to date attempting to link water quality and cancer risk must be considered preliminary and they have not always given consistent results. Although it is not yet possible to fully evaluate and quantify the health hazard (increased cancer mortality) due to chlorination of drinking water, it is thought that there may be no "safe" or "no-effect" levels for organohalogens. Besides the estimates on health risks from chloroform, knowledge is still lacking on the potential hazards from the large number of unidentified organohalogenated compounds encountered in water. Thus prudence is required and it is justifiable to maintain organochlorine concentration as low as feasible in drinking water supplies.

C. TREATMENT AND DISINFECTION PROCESSES CONCERNED

Over the past few decades, potable water supply has generally been characterised by 1) a net decrease in the

174

quality of many raw waters used and 2) the consequent intensification of the treatment applied. The parallel increase of organic pollutants in waters and chlorine levels used in treatment (such as "break-point" chlorination) has led to high organohalogen concentrations in a number of drinking water supplies.

The current practice in many drinking water treatment plants is to use chlorine extensively throughout the system, including raw water transportation, purification treatment, disinfection, and distribution networks. The use of chlorine corresponds to specific functions in these four stages, but organohalogen formation will take place all along the system. Any realistic control policy should consider carefully these stages:

a) Raw water transportation: Chlorine is often used here for its biocidal effect, i.e. to prevent growth of fixed organisms in the mains. Other techniques can be used such as preliminary filtration and clarification of raw waters before transportation, mechanical cleaning etc.

b) Purification treatment: Oxidants are used here for several purposes:

   i) the oxidant effect is mostly aimed at removing various organic and inorganic contaminants such as ammonia and substances causing taste, odour or colour. "Break-point" chlorination is frequently carried out to remove ammonia but various alternatives can be applied such as biological removal, storage in reservoirs or ion exchange. Colour can often be effectively removed by coagulation, and powdered activated carbon dosing is usually effective for taste and odour control.

   ii) the biocidal effect is used in different parts of the treatment (filters, settling tanks, etc.) to prevent growth of algae and other fixed organisms. High chlorine doses may be used, especially in Summer, for this purpose. Various alternatives (physical, mechanical or chemical) exist for this application.

   iii) other miscellaneous effects, such as action on colloids and sludge, can be replaced by other approaches.

c) Disinfectant effect: This is the primary purpose for which chlorine and other oxidants are being used:

i) for waters drawn from polluted sources, disinfection is necessary. Various approaches exist: the use of an oxidant such as chlorine, ozone, chlorine dioxide etc. or UV treatment. Various filtration techniques such as: slow sand filtration, bankside filtration, surface infiltration (on soil or dunes) are very effective and substantially reduce the need for disinfection;

ii) for waters drawn from unpolluted and well protected sources (this is the case for ground waters especially); different viewpoints exist however in OECD countries:
   - in a number of countries it is judged that systematic "chemical" disinfection is unnecessary in the case of waters of good biological quality and is thus often not applied;
   - in some other countries, however, regular chemical disinfection is applied to such waters.

d) Residual "bacteriostatic" effect in the distribution network: This is also a controversial point within OECD countries. In some countries, it is common practice to maintain a chlorine residual in the distribution networks. Besides chlorine, other persistent disinfectants can also be used such as chlorine dioxide or chloramines. In other countries, however, it is not common practice to maintain a chlorine residual in the distribution network under normal conditions, although this may be done on some occasions (when a network is not in a good state of maintenance for instance). A clean and well maintained network is finally a most recommendable policy which may help to minimise application of final disinfectant.

The different oxidants used as treatment reagents and disinfectants have advantages and disadvantages, both in terms of their effectiveness and the by-products they may generate. The alternatives to chlorine, which have already been used in full-scale operations over a certain period and for which experience exists, are ozone and chlorine dioxide. Ozone has been used for potable water disinfection since the beginning of this century. It is an efficient oxidant and a powerful disinfectant but does not leave a residual in the distribution system; therefore, where necessary, a bacteriostatic (such as chlorine dioxide, chloramine or chlorine) must be added.

Chlorine dioxide is also an efficient oxidant and a very good disinfectant; it leaves, like chlorine, a residual in the distribution system. However, it does not remove ammonia.

Little is known as yet about the possible by-products formed from the use of ozone and chlorine dioxide and there is concern about the chlorite and chlorate generated when using chlorine dioxide as well as the fact that some organohalogens may also be produced. Ozone and chlorine dioxide are more satisfactory for taste and odour problems than chlorine and have in the past been used for this reason.

Problems relating to the use of oxidants are: the difficulty of controlling the dosage which has to be constantly adjusted to the plant flow rate and the oxidant demand; the preparation of the oxidant (if this is carried out in situ); and the problem of storage and transportation of potentially dangerous materials (chlorine). The use of chlorine has been favoured in the past, because it is cheap, an effective oxidant and disinfectant and because it is not difficult to use. The microbiological quality of water is, like its chemical quality, of prime importance and any change in chlorination practice should respect the bacteriological safety of the water. If process changes are made, the frequency and type of monitoring for bacteriological and chemical quality may need to be reviewed.

D. CONTROL STRATEGIES

a) Cost considerations

The cost of water treatment is generally a small fraction of the consumers' cost for drinking water. In many cases minor modifications to existing treatment aimed at minimising the precursors before the application of oxidants and optimising the application of oxidants, without endangering the biological quality of the water, will be effective for little or no cost in substantially reducing the by-products formed. As the cost involved is usually moderate, it is prudent to carry out these modifications where practically feasible.

Using alternative oxidants to chlorine will result in somewhat increased water treatment costs although still only a very small and probably negligible increase in the total cost of water supplied. The use of chloramines would not appreciably alter costs; ozone would require substantially increased capital expenditures (compared to chlorine); whereas chlorine dioxide would need lower initial capital cost.

Use of certain treatments such as granular activated carbon or resins to remove organochlorine by-products after formation would be by far the most costly option, and probably only needs to be considered in those cases where water quality is so poor that other conventional technologies cannot be optimised to reduce sufficiently the oxidant demand and precursors.

Control options available to small water systems may be considerably different to those of large systems because small systems have higher per capita costs, less access to trained operating personnel and less capacity to perform sufficient operational monitoring. The use of high quality raw waters is thus particularly important in this case as it makes the whole treatment and distribution far easier and safer.

b) Control options

There are two general approaches to control the organohalogens in the drinking water: a) to prevent their formation or b) to remove them after they have been formed. The latter approach b) is not at present as cost effective as preventing their formation, since organochlorines, once formed, are generally very persistent products which pass through conventional treatment (requiring granular activated carbon or resins for lowering their level). The preventive approach a) is the safer and the better method and in general may be achieved in the following ways:

1. By encouraging, as a fundamental measure, the selective use of raw waters of better quality. Where this is feasible the use of chlorine (or other oxidants) can be avoided or at least minimised.
2. By using alternative purification processes (filtration, precipitation etc.) which permit to minimise in some countries and in others to avoid the use of chlorine or other oxidants at any stage. This approach is particularly advisable in the case of raw waters which are moderately or not polluted.
3. By minimising the dose of chlorine applied and limiting its use to final disinfection only. This approach may be practical in many situations (small water supply installations for instance) and will be made easier if combined with the alternative processes considered in 2) above. Other oxidants can also be used.
4. By minimising the organic precursors as much as feasible before any chlorine application is made (at the very end of treatment). This is a basic

178

approach for raw waters of mediocre quality where both the "precursor" content and chlorine application may be substantial.

5. By carefully controlling the conditions of (i) raw water transportation and (ii) potable water distribution, as these may be major sources of organochlorines in drinking water:

- chlorination of raw waters (a neglected but frequently important organochlorine source) should be avoided and replaced by alternative approaches (clarification of water before transportation, mechanical cleaning etc.);

- when and where a chlorine residual is judged necessary in the distribution network, it should be kept as low as possible; less reactive oxidants such as chloramines or chlorine dioxide may also be used. Good maintenance and cleanliness of the networks are of great importance; they contribute to biological safety and help to minimise the formation of organochlorines (through lower dosing of chlorine residual and lower organic content in the pipes).

The careful and moderate use of chlorine is not condemned, but more prudence and selectivity(*) are required in its utilisation as there is concern about operating practices that are not cognisant of the problems of by-products and do not attempt to minimise them. Obtaining chemically and biologically safe water remains the central goal of drinking water supplies in OECD countries.

---

*) The use of chlorine should be limited to final disinfection; all its other uses in drinking water preparation should preferably be replaced by the other available processes, free of toxic effects.

IV.2. ORGANOCHLORINE COMPOUNDS ENCOUNTERED IN DRINKING WATER

A. TYPES (Volatile and Non-volatile)

The total quantity of organochlorinated compounds in water can be estimated from the analytical determination of the total organic chlorine (TOCl) content of the water. Only a minor part of the compounds measured by the TOCl content can be easily identified. These compounds can be grouped into two broad categories (see Figure A).

a) Volatile compounds (products of low molecular weight)

These compounds may average between 5 per cent to 10 per cent of total organohalogens in raw water and between 15 per cent to 25 per cent in treated waters. The group of trihalomethanes (THMs) (essentially chloroform, but also monobromodichloromethane, monochlorodibromomethane, and more rarely bromoform) constitutes a major part of the volatiles, especially in treated waters. The other volatile compounds, which are mostly industrial solvents, include trichlorethylene, dichloroethane, carbon tetrachloride etc.

b) Non-volatile compounds (products of higher molecular weight)

These products may average between 90 per cent to 95 per cent of total organohalogens in raw waters and between 75 per cent to 85 per cent in treated waters:

    i) the identified part generally constitutes less than 1 per cent of non-volatile organochlorines but includes products such as, pesticides, PCBs and chlorophenols, which may be biologically very active.

   ii) the non-identified part generally constitutes 99 per cent of non-volatiles, and more than three-quarters of all organochlorinated compounds present in raw drinking waters. It

Figure A

TYPICAL DISTRIBUTION OF ORGANOHALOGENS IN WATERS
IN INDUSTRIALISED COUNTRIES

This figure illustrates a typical distribution of organohalogenated compounds that may be found in raw waters drawn from polluted rivers in industrialised countries and in treated waters (after chlorine treatment).

RAW WATERS :
RANGE TOCl = 10 to 100 μg/l
RANGE THM = 2 to 20 μg/l

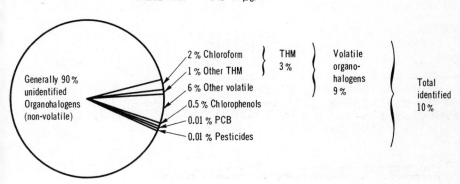

DRINKING WATERS :
(after treatment)
RANGE TOCl = 50 to 500 μg/l
RANGE THM = 10 to 100 μg/l

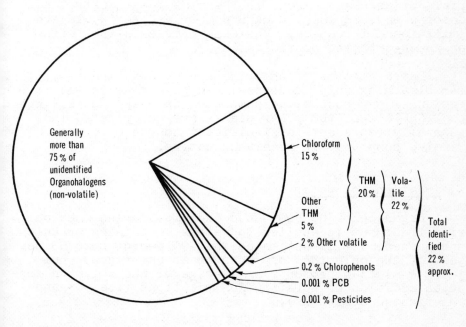

probably includes a large number of different chlorinated hydrocarbons, chlorinated polyphenol etc. The percentages mentioned here can vary widely.

A number of these organohalogenated compounds, whether volatile or non-volatile organochlorines, are highly resistant to biodegradation and can accumulate either in sediments or in the biomass. The more they are halogenated, the more they become lipophilic and can be regarded as suspect or potentially dangerous, even at very low concentrations in the case of long-term exposure

B. ORIGIN

Organochlorine compounds in drinking water have two possible origins: (1) industrial products (or by-products) present as pollutants in raw waters, and (2) drinking water treatment itself.

a) Industrial Products and By-products

Most of these substances do not occur naturally and are products used as solvents, plasticisers, pesticides etc. Chlorinated biphenyls, for instance, have been extensively used in industry (either as dielectric materials or as plasticisers in the plastics industry) but despite bans imposed in OECD countries, residuals are still found in waters. The same problem arises with organochlorinated pesticides.

Large amounts of chlorine are used by industry (i.e. for bleaching pulp and paper) and give rise to the formation of organochlorine compounds as by-products. When cooling waters from power plants are chlorinated this generally represents a significant source of organochlorinated compounds in receiving waters.

b) Treatment of Water

Organochlorine compounds are also formed when chlorine treatment is applied to waters containing organic substances. It is estimated that 95 per cent to 99 per cent of the chlorine dose ends up as chlorides. But 1 per cent to 5 per cent on average of the chlorine produces addition or substitution reactions (i.e. organochlorinated compounds).

i) Disinfecting waste water with chlorine, as practised in a few Member countries, leads to the
formation of chlorinated substances which will
be found in raw waters used downstream for
drinking water supply.

ii) Treatment of drinking water with chlorine began
some 80 years ago. In 1974, however, ROOK (1)
and BELLAR (2) identified haloforms for the
first time in chlorinated drinking water, the
most common being chloroform (3). These compounds had not been identified earlier probably
because the available techniques were not adequate to determine organochlorine compounds by
solvent extraction and gas chromatography. As
early as 1922, DONALDSON (4) was convinced that
chlorine reacted with organic matter to produce
chlorinated compounds. Other authors have also
mentioned this, especially in relation to taste
of the water (5-8).

C. ORGANOHALOGENS ALREADY PRESENT IN RAW WATER
(untreated)

a) Surface Waters

i) Volatile compounds. Numerous studies of surface
waters in various Member countries have shown
that these contain volatile organochlorinated
compounds (Figure A) due to pollution from industrial and urban origin. For example, Rook
et al in Holland (9), Bush in New York (10),
Kühn and Sanders (11), Bohn (12) in Germany and
Switzerland, and Montiel in France (13) have
reported such observations on raw waters. The
results are given in Tables 1 (A, B, C, D, E)
and Table 2 (A, B, C).

ii) Non-volatile compounds are the major part and
include chlorophenols, chlorobiphenyls, chloro-
ligno-sulphonates, pesticides etc. as well as a
large number of non-identified compounds.
After the banning or restriction of the use of
certain organochlorine pesticides in OECD countries, current concentrations in water are now
down to some nanogrammes per litre.

Table 1.A. FRANCE

LEVELS OF VOLATILE ORGANOHALOGENS AT TWO SITES ON THE SEINE (1977)

Site	$CHCl_3$ µg/l	$CHCl_2Br$/ µg/l	$CHClBr_2$/ µg/l	$CHBr_3$/ µg/l	CCl_4/ µg/l
Orly	6.84	3.18	3.36	0.80	0.20
Ivry	4.08	2.63	9.86	0.97	0.22

LEVELS OF VOLATILE ORGANOHALOGENS IN MARNE WATER (St. Maur) BEFORE AND AFTER CHLORINATION (10 ppm Cl_2) (1977)

	$CHCl_3$ µg/l	$CHCl_2Br$/ µg/l	$CHClBr_2$/ µg/l	$CHBr_3$/ µg/l	CCl_4/ µg/l
Before Treatment	2.61	3.02	1.78	0.84	0.17
After Treatment	57.41	18.72	11.86	0.5	0.56
Increase	(x 21.99)	(x 6.19)	(x 6.66)	(x 0.59)	(x 3.29)

Table 1.B.

LEVELS OF VOLATILE ORGANOHALOGENS IN SURFACE WATERS IN GERMANY AND SWITZERLAND

Waters / µg/l	$CHCl_3$	$CHBrCl_2$	$CHBr_2Cl$	$CHBr_3$	CCl_4	C_2HCl_3	C_2Cl_4	$C_2H_2Cl_4$	TOC [mgC/l]
Rhine (Duisburg 15.9)	2.3	0.2	0.3	0.4	0.6	0.7	0.5	0.1	4.2
Main (Frankfurt 14.9)	3.9	<0.1	0.1	<0.1	n.n.	2.4	1.8	n.n.	6.9
Ruhr (Malingen 6.1)	3.3	n.n.	n.n.	n.n.	0.1	2.0	1.8	n.n.	3.0
Neckar (Stuttgart 20.9)	0.5	0.4	0.4	0.5	0.3	0.7	2.0	0.3	4.2
Donau (Langenau 18.1)	1.6	0.7	1.5	n.n.	0.3	0.7	2.0	n.n.	3.3
Bielersee (5.9.77)	0.2	n.n.	n.n.	n.n.	0.02	0.2	0.1	n.n.	2.1
Bodensee (Konstanz 7.9)	0.2	0.2	0.3	0.4	0.03	0.4	0.1	0.2	1.5
Bodensee (Kreuzlingen 6.9)	<0.1	n.n.	n.n.	n.n.	0.01	0.1	0.04	n.n.	1.4
Zürichsee (Lengg 6.9)	0.1	n.n.	n.n.	n.n.	0.01	0.2	0.04	n.n.	1.5

Table 1.C.

LEVELS OF VOLATILE ORGANOHALOGENS IN THE RIVER RHINE

Place / µg/l	$CHCl_3$	$CHBrCl_2$	$CHBr_2Cl$	$CHBr_3$	CCl_4	C_2HCl_3	C_2Cl_4	$C_2H_2Cl_4$	TOC [mgC/l]
Basel	0.9	0.1	0.1	0.2	0.03	3.1	1.0	n.n.	2.2
Mainz	2.2	<0.1	0.1	0.2	0.03	3.7	0.6	0.2	3.9
Koblenz	4.7	n.n.	n.n.	n.n.	6.2	1.0	1.1	n.n.	4.3
Düsseldorf-Benrath	2.9	<0.1	n.n.	<0.1	1.7	0.8	0.9	n.n.	4.2
Düsseldorf-Filhe	3.9	<0.1	n.n.	n.n.	2.4	0.8	1.1	n.n.	4.7
Duisburg-Wittlaer	2.3	0.2	0.3	0.5	0.6	0.7	0.5	0.4	4.2
Duisburg-Hanborn	1.2	0.4	0.5	0.5	0.2	0.8	0.3	0.3	3.8

Table 1.D.

LEVELS OF VOLATILE ORGANOHALOGENS IN THE RIVERS RHINE, LIPPE AND RUHR

River / µg/l	$CHCl_3$	CCl_4	$CHCl=CCl_2$	$CCl_2=CCl_2$
Rhine	0.01-28	0.2-2.7	0.1-2.4	0.2-1.3
Lippe	3.8 -16.4	0.2-6.8	2.1-4.1	1.5-3.2
Ruhr	0.2 -12	0.1-0.7	0.3-0.9	0.4-1.0

185

Table 1.E.

WATER ANALYSES

LEVELS OF VOLATILE ORGANOCHLORINES (OTHER THAN HALOFORMS) IN
SOME EUROPEAN SURFACE WATERS

Site	Country	Concentrations ($\mu g/l^{-1}$ = ppb)		
		$CCl_2 = CHCl$ Trichloro-ethylene	$CCl_2 = CCl_2$ Tetrachloro-ethylene	$CCl_3 - CH_3$ 1.1.1.-Tri-chloroethane
Liverpool Bay	G.B.	0.3 - 3.6	0.12 - 2.6	$\leqslant$0.25 $\leqslant$3.3
Rhine	D			
- Hoenningen		1.0 - 1.5	1.5	
- Luelsdorf		2.0 - 2.5	2.0 - 2.5	
- Wesseling		1.5 - 2.0		
Salzach				
- Marienberg		0.4 - 2.1	0.6 - 1.9	
- Uberackern		25.0 - 73.9	3.3 - 19.6	
Isar				
- At its source		0.02 - 0.03	0.01 - 0.02	
- Munich		0.2 - 0.6	0.1 - 1.0	
- Munich (down-stream)		2.5 - 3.2	1.9 - 2.5	
- Lake Starnberg		0.13 - 0.15	0.15 - 0.20	
- Lake Lerchenau		3.2 - 8.5	2.0 - 2.8	
Twente Canal Hengelo	N.L.	0.26	0.3	0.07
Twente Canal Delden		< 0.2	< 0.2	< 0.1
Eems		11.0	16.0	< 0.1
Oostfriese Gaatje (south)		7.5	6.6	0.3
Oostfriese Gaatje (north)		0.7	1.4	0.1
Ranselgat		0.2	1.7	0.3
Hulbertgat		< 0.2	1.4	0.2
Durance	F			
- Pont Oraison		6.0 - 25.0	<10.0 - 46.0	Undetected
- Ste Tulle		$\leqslant$ 3.0 - 9.0	$\leqslant$ 5.0	Undetected

Table 1.E. gives the trichloroethylene, tetrachloroethane and
1.1.1.-trichloroethane contents of surface water in several
European countries.

Table 2.A. - GERMANY

EFFECT OF CHLORINATION ON NECKAR WATER

After - Chlorination with 20 mg Cl_2/l Floculation with 5 mg/l Al^{3+}			
Date	$CHCl_3$	$CHBrCl_2$	TOCl µg/l
15.6 to 27.9	34	11	677

Table 2.B.

COMPARISON OF VOLATILE ORGANOHALOGEN LEVELS IN
DRINKING WATER

(Germany and U.S.A.)

Substances	Federal Republic of Germany		U.S.A.
	Average	Minima - Maxima (nanogrammes)	Minima - Maxima (nanogrammes)
CH_2Cl_2	846	< 80 - 9,000	
$CHCl_3$	2,457	< 10 - 48,000	200 - 311,000
CCl_4	35	< 0.1 - 500	2,000 - 3,000
C_2Cl_4	721	< 1 - 13,000	200 - 6,000

Table 2.C.

EFFECT OF CHLORINATION ON THE OCCURRENCE OF SOME
HALOGENATED COMPOUNDS IN TAP WATER CONCENTRATION
RANGE in µg/l (Data from the Netherlands)

Parameters	Type of Treatment with Chlorine		
	None	Final Disinfection only	Break-point Chlorination
Number of supplies studied	13	4	3
Chloroform	0.01-2.0	0.1 - 10	25 - 60
Bromodichloromethane		0.01 - 10	15 - 55
Dibromochloromethane		0.01 - 5	3 - 10
Dichloroiodomethane		0.01 - 0.3	0.01 - 10
Bromochloroiodomethane		0.01 - 0.03	0.01 - 0.3
Bromoform		0.01 - 1.0	3.0 - 10
1,1-Dichloroacetone		0.005	0.1 - 1.0
Trichloronitromethane		0.01 - 3.0	0.01 - 3.0

This table shows that, compared to the same treated waters
receiving no chlorine application, a chlorine disinfection
only may increase the organochlorinated compounds by 10 times
or more, and break-point chlorination by 100 times or more.

b) Groundwater

These waters are generally of good quality(*) if correct protection measures have been taken. However, in some groundwaters, organochlorinated pollutants such as trichloroethylenes have been found, due to pollution from spillages or refuse dumps. These compounds are stable and seep into the ground almost ten times faster than water. Such pollution can therefore appear prior to the standard indicators of water pollution in groundwaters (i.e. chlorides).

D. ORGANOHALOGENS CREATED DURING DRINKING WATER TREATMENT

Monitoring of the volatile organohalogen compound content in chlorinated drinking water has shown that:

- surface waters receiving chlorination treatment generally produce organohalogenated compounds;
- properly protected groundwaters, when chlorinated, rarely give rise to significant organohalogen levels unless they contain humic acids (under certain geographical conditions).

This was observed in a number of countries by BUSH (10), ROOK (1), PIET (14), KUHN (3), BOHN (12) and MONTIEL (13).

E. ORGANOHALOGENS FOUND IN TAP WATER

The water supply system can interfere with the quality of water delivered to the consumer. In the distribution system, the formation of additional organohalogens and a change in their composition can continue as long as free chlorine and organic compounds are both present in the water.

In PVC pipes used for distributing water, vinyl chloride monomers have been detected at levels of 1 µg/l (or less) but exceptionally up to 10 µg/l. This substance may be released over a long period, decreasing with time. Moreover, when the grounds, through which

---

*) However, in an increasing number of aquifers, the high nitrate content is now a very serious problem.

PVC pipes are laid, are contaminated with organochlorines
(e.g. pesticides), they can seep through the PVC pipes
into the water.

SUMMARY

Organohalogenated compounds (both volatile and non-
volatile) may already be present in polluted surface
waters and in poorly protected aquifers; but their concentration can increase considerably following chlorination treatment if organic precursors are present.

However, some waters, which have not been subjected
to man-made pollution may also lead, after chlorination,
to the formation of high amounts of organohalogens. This
occurs particularly when treating water containing high
amounts of precursors such as humic and fulvic acids, or
algae metabolites from eutrophic waters.

IV.3. <u>POTENTIAL TOXICITY OF ORGANOCHLORINE COMPOUNDS</u>

A. INTRODUCTION

Among the halogenated compounds in drinking water, most of the attention of health authorities has been given to the trihalomethanes, as this group of compounds is both the easiest to identify and the main group individually (about 20 per cent quantitatively of total organohalogens) found in drinking water.

Within this group of trihalomethanes, chloroform is generally the main constituent; this is why toxicological studies on drinking water chlorination have focused particularly on chloroform. It should, however, be realised that:

a) chloroform generally represents only about 15 per cent of total organohalogens;
b) among the remaining part and especially the unidentified compounds (80 per cent of the total), many substances may present potential health risks at least similar to chloroform or greater;
c) unknown synergistic (*) effects, which have not yet been evaluated, may occur between organohalogens and other contaminants present in water.

For these reasons the data relating to chloroform, as discussed below, only cover part of the health risks involved. It has been suggested, however that the magnitude of the maximum health risk estimated for chloroform might also give an indication of the possible health risks related to water chlorination in general, but great doubts are expressed about the validity of this extrapolation.

――――――――――

*) The potential for synergistic effects between organohalogens and trace pollutants cannot be disregarded, Chloroform, which for instance is the solvent used in analytical chemistry to extract organic pollutants from waters, may also encourage the absorption of a number of trace pollutants in the human organism. At present this is only a hypothesis and very little is known about this process.

B. CHRONIC TOXICITY (INCLUDING CARCINOGENICITY) OF CHLOROFORM

Except for those on the carcinogenicity of chloroform, few studies of the chronic toxicity of chloroform have been carried out. The long-term oral administration of chloroform at a dose of 0.4 mg/kg body weight did not produce any changes in the investigated indices in albino rats, and the only effect of this dose on guinea pigs was an increase of vitamin C in the adrenals (15). The safety of adding chloroform to toothpaste and mouthrinse was assessed in two long-term studies, involving 229 human subjects (16). Daily consumption of chloroform was estimated to be 0.34 to 0.96 mg/kg over a 1- to 5- year period. Results from this study showed no hepatotoxicity based on liver function tests. Reversible hepatotoxicity was the only effect observed in a 47-year-old man who every day for ten years had consumed between 12 and 20 ounces of a chloroform-containing cough suppressant; his daily dose of chloroform was estimated to be between 23 and 37 mg/kg body weight (17).

In an investigation carried out to examine the safety of chloroform-containing toothpaste (18), an excess of adenomatous tumours of the renal cortex were found in male mice (ICI Switzerland) receiving a daily dose of 60 mg chloroform per kg body weight over a 100-week period. No tumours were reported in animals consuming a daily chloroform dose of 17 mg/kg body weight. Even at the higher dose, an excess of tumours was not found in female mice (ICI Switzerland), in three other strains of mice, or in rats. Negative results were also obtained with beagle dogs; but this part of the study may have been too short.

Note: <u>Acute Toxicity of Chloroform</u>

Chloroform is a central nervous system depressant. It also affects the liver and kidney. The acute toxicity of chloroform has been studies in many species. Reported LD_{50}s in mice range from 118 to 3,286 mg/kg body weight, depending on the route of administration (13,10). Oral LD_{50}s for chloroform in rats range from 419 mg/kg to 1,257 mg/kg. The mean lethal dose in man is considered to be about 44 g or 630 mg/kg body weight for a 70 kg man; however, ingestion of more than 250 g of chloroform has been survived (12). The lowest published lethal dose of chloroform in man is 210 mg/kg body weight (2).

In another study conducted by the U.S. National Cancer Institute (NCI) (19), chloroform dissolved in corn oil was administered by gavage to Osborne-Mendel rats and $B_6C_3F_1$ mice at two dose levels five times per week. Dose levels of 90 or 180 mg/kg body weight were given to the male rats for 78 weeks; the female rats received doses of 125 or 250 mg/kg for the first 22 weeks and the same dose as the males thereafter. After 111 weeks the rats were sacrificed and a statistically significant incidence of kidney epithelial tumours was found in the males (24 per cent in the high-dose and 8 per cent in the low-dose groups) but not in the females. The male mice first received doses of 100 or 200 mg/kg, and the females were initially dosed with 200 or 400 mg/kg. After 18 weeks, the doses were changed to 150 and 300 mg/kg for males and to 250 and 500 mg/kg for females. Highly significant increases in hepatocellular carcinoma were found in both sexes: 98 per cent of the males and 95 per cent of the females at the high dose and 36 per cent of the males and 80 per cent of the females at the low dose. It should be emphasized, however, that the doses used in the NCI studies are extremely high and, as a greater than 10 per cent weight loss was observed in the animals, such doses can be considered to be higher than a true maximum tolerated dose.

Long-term toxicity tests were carried out in France (35) recently (mice and rats) using concentrated chloroform extracts of micropollutants from chlorinated drinking water prepared from surface water sources. They showed a significant increase in the incidence of various types of malignant tumours and in general a significant increase in mortality rates (especially males). Similar tests carried out with extracts from non-chlorinated groundwaters showed no ill-effects.

C. MUTAGENICITY STUDIES RELATING TO WATER CHLORINATION

At present a number of studies of the mutagenicity of (concentrates of) mixtures of organhalogenated compounds found in drinking water are being carried out in several OECD countries. Mutagenicity of drinking water can be detected by different types of mutagenicity screening tests, such as the Ames test, where chlorination has been applied to polluted raw water. Although such information may help to identify the most hazardous contaminants in the raw water, mutagenicity tests cannot provide data on which quantification of the potential health risk for the population can be based. For this purpose, chronic toxicity data and epidemiological studies are essential.

D. EPIDEMIOLOGY RELATING TO DRINKING WATER CHLORINATION

Epidemiological studies have several real advantages over experiments with laboratory animals in the assessment of the effects on humans of low-level exposure to environmental contaminants. Observations can be made directly on human populations, and extrapolation from unrealistically high exposures is not required. On the other hand, the epidemiological approach is limited by the lack of a systematic data base on organic contaminants in drinking water; available data are usually of recent origin and may not reflect exposure levels at the time (possibly decades earlier) a cancer was induced.

Recent investigations (21) have shown that cancer rates were higher in Louisiana parishes that used the Mississippi as a water source. It was also found that certain substances (tetrachloroethylene and carbon tetrachloride) were not only present in the drinking water of New Orleans but in the blood of people who consumed the water (1).

In related studies of counties using water from the Ohio River (2), it was concluded that the relationships found in Louisiana were not always evident in the cancer death rates of Cincinnati or the Ohio River counties; however, the cancer rates for all sites combined were elevated in areas with surface water sources. A second finding in this investigation was a positive correlation between the practice of prechlorination and cancer incidence. Prechlorination is linked with increased organochlorine production, but it is not yet clear whether the apparently higher cancer rates, where surface waters are consumed, are only due to organochlorines or also to other pollutants. The same question has been posed in other epidemiological studies carried out in Member countries and it may reasonably be assumed that both factors contribute to increased mortality rates.

In the U.S. National Organics Reconnaissance Survey (NORS) (3), a study was made of the chemical concentrations and death rates in the communities served by the water supplies sampled. A significant relationship was reported between the chloroform concentration and the total death rate from cancer.

In the Netherlands an epidemiological study (Kool c.s., 1979) carried out on one-third of the population has shown that in areas where drinking water is prepared from surface waters of poor quality (with chlorination treatment) a substantially higher cancer mortality rate (+ 6.7 per cent) is found (especially for cancers of the esophagus and stomach) compared to areas where it is prepared from good quality ground waters (generally with no chlorination treatment). Drinking water receiving chlorine disinfection was found to contain an average of ten

times more organohalogenated compounds than that receiving
no chlorination treatment; water receiving prechlorination
or break-point chlorination had an organohalogen content
more than 100 times greater. Besides the trihalomethanes
other individual organic compounds like chlorinated ben-
zenes and anilines, originating from the raw water, also
showed statistical correlations with cancer mortality.

The results of such epidemiological studies must be
interpreted with caution. A positive correlation obtained
by a retrospective epidemiological study cannot establish
with certainty the direct causality because of the many
known or uncontrollable factors that are not taken into
account. Reliable estimates of the carcinogenic health
risk resulting from the presence in drinking water of organics in general and organochlorines in particular will
not be really available until further investigations are
carried out.

E.   TENTATIVE MODELS FOR ASSESSING HEALTH RISKS

In the light of the limited epidemiological data, a
conservative approach in assessing the risk from organochlorines in drinking water is to assume that man responds
to chloroform in a similar manner to the rodent in terms
of hepatotoxicity and carcinogenesis and, as reflected by
the NCI results with the Osborne-Mendel rat (19); the kidney is the target organ.

Extrapolation of results obtained in rodents exposed
to chloroform in quantities above the maximum tolerated
dose to their effects in humans consuming organochlorines
at the low levels that occur in drinking water should,
however, still be regarded as tenuous.

The potential of organic contaminants, such as organohalogens, in drinking water to affect human health adversely cannot be assessed with certainty at present.
The key issue is whether the organochlorines at low concentrations can cause or increase the incidence of cancer
in man. Since chloroform is the principal trihalomethane
in drinking water and very few data on the toxicity of
the other organochlorines are available, it is assumed
that the risk from exposure to any of the trihalomethanes
and other organochlorines is equivalent to the risk from
exposure to the same concentration of chloroform. This
theory, however, is purely hypothetical. In order to assess the risk to man from ingestion of chloroform, mathematical or probability models are used to determine the
maximum risk for low-dose exposure. Four such models are
(a) margin of safety, (b) one-hit, (c) two-step, and (d)
probit-logarithm.

194

Determination of Maximum Risk from Chloroform Ingestion based on Rat Kidney-Cancer Data

(Minimum-effect dose = 90 mg/kg/day)

Model	Estimated Maximum Risk	Maximum Daily Dose mg/kg
Probit-log (slope = 1)	$1.6-4.0 \times 10^{-8}$/yr.	0.01
Probit-log (actual slope)	1×10^{-9}/lifetime	0.01
Linear (one-hit)	0.42×10^{-6}/yr.	0.01
Two-step	0.267×10^{-6}/yr.	0.01

Current maximum risk values estimated by the U.S. National Academy of Sciences for Chloroform are 1.7×10^{-6} per g of lifetime daily exposure. This means that the incremental lifetime risk at 100 ug/l (assuming two litres of water consumed per day) computes to $2 \times 1.7 \times 10^{-4}$ or about 3.4 per 10,000 increase in mortality.

Canadian viewpoint on the question is not dissimilar to the United States and risk assessments using such data suggest that low doses of organochlorines in potable water may increase the incidence of some cancers. A number of mathematical models have been used to assess this risk for the population exposed.

It must be stressed, however: 1) that many tenuous assumptions have been made on the mode of action of the chemical and its metabolites (for example, that the metabolism of chloroform in humans is the same as its metabolism in rodents) and 2) that the biological model is used in an equally tenuous mathematical model and that data are only limited to kidney cancer and chloroform (the other organohalogens and the cancers of other organs have been disregarded). Only when the validity of these assumptions has been established and when more data are available will it be possible to describe risk with more confidence by this technique. It is also worth noting that the results from human epidemiological studies, appear to be more serious in their provisional conclusions than suggest the risk assessment made on chloroform and rats.

$$\circ \ {}_\circ \ \circ$$

It can be concluded that the risks of increased cancer mortality due to chlorination of drinking water cannot yet be fully evaluated but it is clear (on the basis of present data), that they are worth considering. In fact, apart from the estimates on health risks from chloroform, knowledge is still lacking on the potential hazard from the large number of unidentified organochlorinated compounds encountered in drinking water; as there may be no "safe" or "no-effect" levels for these substances, prudence is required and it is justifiable to maintain their concentration as low as feasible.

IV.4. <u>REGULATORY MEASURES</u>

A. APPROACHES TO THE QUESTION

Although the problem of organochlorines in drinking water has recently raised concern in various Member countries, very few countries have so far fixed or prepared specific standards. Moreover, for the few countries where this is the case, the standards are far from analogous. It has often been observed that unless a guideline or standard is stringent enough, it may have a disincentive effect on a majority of waterworks which are already probably within the limits fixed. The guidelines or standards should not be the lowest common denominator of the levels attained in the country but should be focussed, with the desirable flexibility, on the best levels realistically attainable.

Procedures for updating levels within a planned timeframe could usefully be provided so that re-evaluation can take place in the light of improved treatment practices, new technologies or medical/scientific knowledge. Provision should also be made for special or difficult cases.

B. NATIONAL STANDARDS AND GUIDELINES

The standards or guidelines in the following counries are:

Germany (FRG): The recommended guideline for THM is 25 µg/l (0.025 mg/l) at consumer's tap (95 per cent average over the year). There is also a standard for maximum final disinfectant concentration: 0.3 mg/l (when water leaves the plant).
United States: The limit for THM is 100 µg/l (0.1 mg/l at consumer's tap (on a 12-month average).
Canada: The guideline for THM is 350 µg/l (0.35 mg/l) as a maximum permissible value; the objective is zero. Monitoring location is understood, of course, to be at the consumer's tap.
Although the present health evidence does not appear to indicate the setting of a more stringent maximum concentration, it is recommended that THMs be

reduced where economically practicable, while maintaining a measurable residual disinfectant power.
This and other drinking water guidelines are reviewed periodically in Canada, and are modified when
new scientific evidence is produced which demon-
strate that current guidelines are inadequate to protect health. The type of disinfecting agent is not
specified in Canada and, although chlorine is used
widely, other disinfectants have been used
successfully.
A number of European countries are envisaging the
future adoption of guidelines.
In the United Kingdom it is not felt that the present health evidence so far justifies the setting of
a standard but, nevertheless, it is advised that THM
levels should be reduced where economically practicable and consistent with the maintenance of bacteriological safety.

IV.5. <u>MECHANISM OF ORGANOHALOGEN FORMATION</u>

A. ORGANIC COMPOUNDS RESPONSIBLE FOR ORGANOHALOGEN
FORMATION

A large range of organic compounds may react with
chlorine (or other halogens) to produce different categories of organochlorinated compounds - either volatile compounds like chloroform and other trihalomethanes or non-
volatile compounds which represent the majority but are,
as yet, very poorly identified. The organic compounds responsible for organohalogen formation or "precursors" may
give rise to one or more types of organochlorines with a
certain "yield".

In the absence of data on the majority of organochlorinated compounds (non-volatile), reference will only be
made to haloforms (volatile).

Generally speaking, humic and fulvic acids, as well
as algae metabolites, are the largest contributors to THM
formation when chlorine is applied. These compounds are
very often present in surface waters and generally constitute more than 80 per cent of the total THM precursors.

For the other "precursors" of industrial origin it
has been shown, for instance, that:

- alcohols produce halomethanes if oxidised to a
  chloroform-yielding molecule;
- methylketones give haloforms with a yield of about
  5 per cent;
- phenols produce haloforms with a yield of about
  5 per cent;
- diphenols and triphenols give haloforms when in
  meta position (resorcinol) with a high yield;
- the other phenol compounds are also haloform precursors;
- β diketones give a high yield of chloroform;
- amino-acids can produce chloroform.

There is another type of compound responsible for
THM formation - synthetic polyelectrolytes. Haloform con-
concentration due to polyelectrolytes is, however, quite
low since these chemicals are used in very small doses.
The use of these products for potable water production is
either restricted or forbidden in several countries.

B. REACTION KINETICS AND MECHANISMS

The main parameters influencing the formation of haloforms are:

- precursor concentration and chlorine dose;
- contact time;
- pH and temperature;
- the nature of precursors present.

Various mechanisms have been proposed depending upon the nature of precursors present. The reaction mechanisms are obtained with the chief primary precursors, namely resorcinol, phenol, methylketones and β diketones.

Secondary precursors are compounds leading to these primary precursors when oxidised.

The importance of pH should be emphasized as it affects the balance and yield of non-volatile/volatile organohalogens.

VI.6. USE OF CHLORINE IN POTABLE WATER TREATMENT

Chlorine was first used because of its disinfecting properties, but since then, it has been increasingly used for many other purposes.

A. CHLORINATION FOR DISINFECTION

Chlorine has two effects:

- it is a disinfectant for water normally used at the end of treatment;
- because of its residual bacteriostatic properties, it prevents bacterial proliferation or contamination in the network (broken pipes, maintenance work, and when connecting buildings to the supply).

a) Bactericidal effect of chlorine

It has been established that chlorine does not act on bacteria by oxidation, instead it inhibits enzymatic cell functions.

The mechanism for sterilising microbes is therefore governed by the diffusion rate of the oxidising molecules through the cell membrane and by the change in the reacttion rate of the cell metabolism caused by inhibition of the enzymes within the cell. The first reaction therefore depends on the forms in which chlorine occurs in water, the temperature of the solution and the sterilant dose.

The hypochlorous form (HOCl) alone is able to cross the cell membrane, hence the link between chlorine's bactericidal effect and pH;

- at pH 7, 80 per cent is in the form of HOCl
- at pH 7.5, 50 per cent is in the form of HOCl

At pH 8 chlorine has little sterilising effect.

i) Penetration into bacterial cells

This depends on the bacterial species concerned, as shown in Figure B.

201

ii) Doses of chlorine

> Actual exposures of 10-20 minutes are generally
> sufficient to obtain a disinfecting effect with
> a free chlorine content of 0.1 to 0.2 ppm at
> pH 7 ($HOCl + HOCl^-$).

b) Virucidal effect of chlorine

Chlorine is unable to act on viruses in the same way
as on bacteria. Numerous studies have now firmly established that the only chlorine compound which can be considered as an active inhibitor of enteroviruses is non-
dissociated hypochlorous acid (22). Here also chlorine
must be in its free form and not form complexes with ammonium ions.

It should also be noted that the absence of faecal
contamination test organisms is not a criterion for treatment effectiveness with respect to viruses (see Figure C).

c) Bacteriostatic effect within the water supply system

A residual of chlorine can be left to prevent bacterial growth and contamination in the supply network. The
bacteriostatic effect is obtained by maintaining a minimum residual of chlorine, i.e. 0.1 mg/l in the supply system. Some countries have standards on chlorine residuals
either at minimum or maximum levels.

B. ROLE OF CHLORINE AS A CHEMICAL REAGENT (Oxidation)

Chlorine is used as an oxidant for colour, taste and
odour removal, ammonia elimination and iron (and sometimes
manganese) precipitation.

In most cases, oxidation occurs but there are also
substitution and addition reactions. The latter are the
most relevant here for example. Organic amines react with
chlorine to give organic chloramines:

$$RNH_2 + Cl_2 \longrightarrow RNHCl + RNCl_2$$

Phenols give chlorophenols after reaction with
chlorine.

a) Ammonia oxidation: The oxidation of ammonia by
chlorine leads to the formation of intermediate byproducts
such as monochloramines, dichloramines, hydroxylamines,
nitroxylamines (22) as well as to the production of nitrogen gas and small amounts of nitrates. The chemical reaction is:

202

Figure B

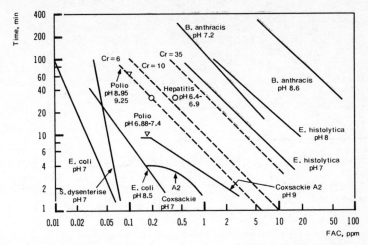

Disinfection versus free available chlorine residuals. (Time scale is for 99.6 to 100 percent kill. Temperature was in the range of 20 to 29°C, with pH as indicated.) Reprinted from Journal American Water Works Association 54 (11), Nov. 1962, by permission of the Association.

Figure C

Comparison of the relative inactivation of poliovirus 1 by hypochlorous acid, hypochlorite ion, monochloramine, dichloramine and chlorine dioxide at 15°C at different pH values. (Journal American Water Works Association.)

$$NH_4^+ + Cl_2 \rightarrow N_2 + HCl + N^+.$$

In theory, 7.6 mg/l of Cl_2 are necessary to oxidise 1 mg/l of NH_4^+ -N. In practice, as much as 8.5 to 9.0 mg/l of Cl_2 are used. The speed of the reaction, and the formation of side reactions like production of nitrogen trichloride, depends strongly on pH and initial ratio of chlorine to ammonia nitrogen.

b) Other oxidation reactions

Chlorine contributes to some oxidation of metals (oxidation of Fe^{++}, Mn^{++}, Cr^{++},) as well as organic molecules (aldehydes, alcohols, sugars, etc.); 95 per cent to 99 per cent of the chlorine added can be assumed to end up as Cl^-, and therefore produces an oxidation reaction.

C. OTHER ROLES OF CHLORINE IN WATER TREATMENT

The paragraphs below discuss the application of chlorine biocidal effects of a technical nature rather than as a disinfectant. However, chlorine application at early stages of the process, in the presence of high precursor levels, may lead to significant organochlorine formation. Moreover, chlorine may prevent useful biological processes (such as biological filtration etc.).

a) Raw water conveying

When conveying raw water over long distances (ten hours or more) chlorine is frequently used to control biological proliferation and to prevent any deposits from fermenting. A dose of 1 or 2 mg/l of chlorine is applied in this case, depending on the chlorine "demand" of the water. However the pipework can be designed so that a brush can be inserted for regular mechanical cleaning, and this solution should obviously be preferred to chlorine addition.

b) Algicidal effect during treatment

The presence of algae in a settling tank can affect the operation of a treatment plant. In summer, photosynthesis releases minute bubbles of oxygen, making the sludge rise to the surface and thus hindering decantation. In the case of lamellated settling tanks proliferation of fixed algae on the lamellae or pipes may obstruct the water flow; the presence of chlorine prevents the algae from multiplying and facilitates the operation of the tank. In general, 0.5 mg/l of chlorine is enough to prevent the harmful algae from developing.

c)   Effect on filtration

It has been established (23) that the growth of organisms on the surface of a filter bed increases the "load loss" and can therefore shorten filtration cycles, resulting in lower outputs. In many cases the micro-organisms concerned have been identified. Some are bacteria, the best known being nitritation and nitration bacteria. Although they increase the filter biomass loss, they may also have a very useful role and help to remove ammonia if the oxygen level in the water is sufficient. The organisms can also be of a larger size and belong to zooplankton (Copepodes). Algae are frequently present, especially diatoms, as they grow throughout the year; the most important are Asterionella, Fragilaria and Synedra (22).

High chlorine doses are often applied to stop the growth of algae and other micro-organisms; this occurs especially during summer when there is both maximum growth of organisms and peak demand for water in many plants. Alternative approaches should be adopted such as shock dosing and rinsing instead of continuous chlorine application to deal with this problem.

d)   Destabilization of colloids

It has been known for many years that chlorine assists water clarification (24). It lowers the turbidity level and often reduces the coagulant dose. It improves the performance of floc blanket sedimentation plants and in some cases lessens the requirement for accurate coagulant dose (amounts above or below the normal dose can be used). Consequently, the flow rate through the decanters can be faster and peak flow operation becomes easier.

The precise role of chlorine here is not exactly known (22). Owing to its oxidising effect, it may render organic complexes insoluble, precipitate iron, reduce algae and bacteria counts, or break down organic molecules responsible for colouring or turbidity (22). However, it is unlikely to have a destabilizing role like a coagulant, which would allow purely inorganic colloids to be removed (25). Even the very slightly lower pH produced by chlorine indirectly aids the removal of organic colloids. Chlorine is frequently used to oxidise ferrous sulphate to ferric sulphate. This process is applied for iron removal as well as to produce a coagulant for removal of suspended matter. An excess of chlorine is used.

D. INTERFERENCE WITH ACTIVATED CARBON

Like all oxidants, chlorine reacts with activated carbon. As a result, it breaks down into hydrochloric acid, and oxide radicals are formed on the surface of the carbon.

Activated carbon therefore acts as a dechlorinator. The reaction kinetics involved have been the subject of much research (10), (11); formulae have been proposed for dechlorination time which depends on the flow rate, the thickness of the activated carbon bed (in other words contact time), the initial chlorine concentration and the type of carbon. The reaction is fairly fast; i.e., 50 per cent dechlorination is achieved within a few seconds. The reaction is much slower with combined chlorine and it is very difficult to break down chloramines with carbon.

The formation of surface oxides (e.g. carboxylic radicals) in the carbon pores appears to hinder adsorption. The organic molecules reacting with chlorine acquire much stronger polarity (12), and overall chlorination appreciably lowers adsorption effectiveness.

Activated carbon granules do not only act through adsorption since much of the treatment effectiveness is due to bacterial activity in the filter. A high chlorine dose added upstream from a carbon filter greatly reduces the beneficial effects of the bacteria (26) in the treatment. Although free chlorine is rapidly broken down, combined chlorine with appreciable bactericidal properties always remains. Nevertheless, chlorination does not seem to completely inhibit microbial activity inside the carbon pores.

Chlorine may help to fix and oxidise organic substances on the carbon (10). The two effects are opposed and therefore it is not clear to what extent chlorination has a negative effect on the efficiency of activated carbon.

IV.7. ALTERNATIVES TO CHLORINATION

Chlorine has many uses in potable water treatment, but alternative technologies are now available. In addition, the need for any treatment involving a chemical oxidant can be considerably reduced by using better quality raw waters and other physical and biological processes.

A. ALTERNATIVES FOR DISINFECTION

The effects of chlorine can be replaced as follows:

a)   Bactericidal effect

This can be obtained with other oxidants such as chlorine dioxide, ozone, chloramines, or to some extent other processes such as UV rays, alkaline treatment at pH12 or certain filtration processes.

b)   Virucidal effect

So far studies have shown that ozone and chlorine dioxide have a definite virucidal effect.

c)   Bacteriostatic residual effect

If a residual disinfectant is desired in the distribution network, chlorine dioxide and chloramines can be used efficiently.

Tables 3 to 5 summarise the characteristics of the major disinfectants.

B. ALTERNATIVES FOR CHEMICAL ACTION (Oxidation)

Here also other processes can be used. Oxidation can be either chemical or biological:

a)   <u>Chemical oxidation</u>

Depending on specific circumstances and on the purpose desired, a specific oxidant will be chosen. Oxidation is used for iron and manganese removal. Raising the
oxygen content of water is sometimes sufficient to oxidise
iron salts or sulphides. A separation step is then required and pH has a strong influence.

In many other instances and particularly for the removal of manganese, a chemical reagent may be required.
Chlorine dioxide, ozone or potassium permanganate can be
used. Chloramines are not sufficiently powerful oxidants
for the purpose.

The chemical oxidants are not all equally effective:
the best one is ozone, closely followed by chlorine
dioxide, whilst potassium permanganate is usually used in
pre-treatment only. However, they can all produce oxidised substances which may be potentially hazardous.

b)   <u>Biological oxidation</u>

Iron and manganese can also be biologically oxidised
as can NH_4^+, NO_2^-, S^-, CN^-. An excess of oxygen, precipi-
tating metal oxides before it reaches the bacteria, may
however, be undesirable.

c)   <u>Removal of ammonia</u>

Ammonia may be removed by oxidation to nitrate, by
conversion to nitrogen (which can only be done with chlorine) or removal of NH_4^+ by ion exchange. Ammonia strip-
ping with air can rarely be efficiently applied in drinking water treatment as ammonia levels are generally too
low.

   i) <u>Biological oxidation of NH_4^+ to NO_3^- (Nitrification)</u>
      This process may be greatly affected by pollution
      (e.g. hydrocarbons or toxic compounds). Further-
      more, some transition metals can either inhibit
      or activate the biological reactions. A minimum
      PO_4^{3-} is essential (0.1 mg/l). Because of the
      limited oxygen content of water, ammonia cannot
      be oxidised beyond 1.5 to 2 mg/l NH_4^+. In order
      to oxidise to higher levels, the water must be
      aerated. The efficiency of this treatment also
      largely depends on the temperature; it decreases
      sharply when temperature goes below 10°C.
      The ammonia biological oxidation stage can be in-
      serted at two different points in the treatment:
      either at the beginning in the form of storage or
      aerated filtration (with the inhibition risks al-
      ready referred to), or at the end of treatment

Table 3

SUMMARY OF MAJOR POSSIBLE DISINFECTION METHODS FOR DRINKING WATER

Disinfection Agent(a)	Technological Status	Efficacy in Demand-free Systems(b)			Persistence of the oxidant in Distribution System
		Bacteria	Viruses	Protozoan Cysts	
Chlorine(c) As hypochlorous acid (HOCL) As hypochlorite ion (OCl⁻)	Widespread use in drinking water in most countries	++++ +++	++++ ++	++ (d)	Good
Ozone(c)	Widespread use in drinking water, particularly in France, Switzerland and Canada (especially in Quebec)	++++	++++	++++	No residual left
Chlorine Dioxide(c),(e)	Widespread use for disinfection (both primary and for distribution system residual) in Europe; limited use in United States to counteract taste and odour problems and to disinfect drinking water	++++	++++	(d)	Good
Iodine As diatomic iodine (I2) As hypoiodous acid (HOI)	No reports of large-scale use in drinking water	++++ ++++	++ ++++	+++ +	
Bromine	Limited use for disinfection	++++(f)	++++(f)	+++(f)	Fair
Chloramines	As a large-scale disinfectant for drinking water their use is presently rather limited	++	+	+	Good

(a) The sequence in which these agents are listed does not constitute a ranking.
(b) Ratings: ++++, excellent biocidal activity; +++, good biocidal activity; ++, moderate biocidal activity; +, low biocidal activity; -, of little or questionable value.
(c) Byproduct production and disinfectant demand are reduced by removal of organics from raw water prior to disinfection.
(d) Either no data reported or only available data where not free from confounding factors, thus rendering them not amenable to comparison with other data.
(e) MCL 1.0 mg/l because of health effects (Symons et al., 1977)
(f) Poor in the presence of organic material.

Table 4

SUMMARY OF MINOR POSSIBLE DISINFECTION METHODS FOR DRINKING WATER

Disinfection Agent[a]	Technological Status	Efficacy in Demand-free Systems[b]			Persistence of Residual in Distribution System
		Bacteria	Viruses	Proto-zoan Cysts	
Ferrate	No reports of use in drinking water	++	++	NDR[c]	Poor
High pH conditions (pH 12-12.5)	No reports of large-scale use in drinking water	+++	+++	NDR[c]	Feasibility restricted as consumption of high pH water not recommended
Hydrogen peroxide	No reports of large-scale use in drinking water	±	±	NDR[c]	Poor
Ionizing radiation	No reports of use in drinking water	++	++	NDR[c]	No residual possible
Potassium permanganate	Limited use for dis-infection	±	NDR[c]	NDR[c]	Good, but aesthetically undesirable
Silver[d]	No reports of large-scale use in drinking water	+	NDR[c]	+	Good, but possible health effects and very expensive
UV light	Use limited to small systems	+++	+++	NDR	No residual possible

(a) The sequence in which these agents are listed does not constitute a ranking.
(b) Ratings: ++++, excellent biocidal activity; +++, good biocidal activity; ++, moderate biocidal activity; +, low biocidal activity; ±, of little or questionable value.
(c) Either no data reported (NDR) or only available data were not free from confounding factors, thus rendering them not amenable for comparison with other data.
(d) MCL 0.05 mg/l because of health effects (Symons et al., 1977).

Table 5

STATUS OF POSSIBLE METHODS FOR DRINKING WATER DISINFECTION

Disinfection Agent	Suitable Inactivating Agent	Limitations	Suitability for Drinking Water Disinfection(a)
Chlorine	Yes	Efficacy decreases with increasing pH; affected by ammonia or organic nitrogen	Yes
Ozone	Yes	On-site generation required; no residual;	Yes
Chlorine dioxide	Yes	On-site generation required; interim MCL 0.5 mg/l (U.S.A.)	Yes
Iodine	Yes	Biocidal activity sensitive to pH	No
Bromine	Yes	Lack of technological experience; activity may be pH sensitive	No
Chloramines	No	Mediocre bactericide, poor virucide	(b)
Ferrate	Yes	Moderate bactericide; good virucide; residual unstable; lack of technological experience	No
High pH conditions	No	Poor biocide	No
Hydrogen peroxide	No	Poor biocide	No
Ionizing radiation	Yes	Lack of technological experience	No
Potassium permanganate	No	Poor biocide	No
Silver	No	Poor biocide; MCl 0.05 mg/l	No
UV light	Yes	Adequate biocide; no residual; use limited by equipment maintenance considerations	No

(a) This evaluation relates solely to the suitability for controlling infectious diseases transmission. See conclusions.
(b) Chloramines may have use as a secondary disinfectant in the distribution system in view of their persistence.

either as biological filtration on sand, volcanic ash, activated carbon or pumice stone support, or using a bacterial fluidised bed. This is a simple and economical method which is recommended as an alternative to break-point chlorination.

ii) Ion exchange resins. Clinoptilolite is a natural ion exchanger which selectively removes ammonium ions in the presence of magnesium, calcium and sodium (unlike cationic synthetic resins, which tend to attract divalent metal cations).
Three different mechanisms may account for the selectivity of clinoptilolite: sieve action, hydration of cations and separation of the anionic sites.
MACLAREN (27) has determined the selectivity coefficient of this exchange resin:

$$K = \frac{cation}{NH_4^+}$$

This technique has the advantage of being almost temperature independent. It may, however, create the problem of regeneration of wastes and it is also rather expensive.

C. ALTERNATIVES TO OTHER ROLES OF CHLORINE TREATMENT

a) Transport of raw waters

Unless raw water is sufficiently pure (for example, certain underground or stream waters), transport of untreated water over long distances may sometimes cause side effects (i.e. deposits of mud, proliferation of organisms) and thus encourage early chlorination. Chlorination of raw waters at this stage can lead to significant organochlorine formation because precursor levels may be high. Various alternative approaches should thus be encouraged:

i) Primary treatment before transport. Primary clarification treatment by various classical processes (related to the quality of raw waters, length of transport, final treatment planned etc.), can be carried out before transportation. The final treatment at the point of use (generally in urban areas) can thus be much simpler and installations can be more compact. This approach is a positive response to the problem.

ii) Mechanical cleaning devices for water pipes are now in common use.

iii) Other oxidants, less reactive than chlorine, such as chloramines or chlorine dioxide can also be used at this stage. However, they may also lead to some byproducts; thus approaches a) and b) seem preferable.

iv) <u>Shock dosing and rinsing</u> (with an oxidant or another chemical)

A high dose of an appropriate reactant is circulated through the section of the network to be cleaned; it is then evacuated and the network rinsed before normal water circulation restarts. This method, at least within its application limits, has the advantage that the reactants are not mixed with the water being transported.

b) <u>Control of algae during treatment</u>

In all waters subject to eutrophication, algae can be inactivated or their growth slowed down by using oxidants, toxic algicidal agents or growth retarders. The oxidants or algicidal chemicals should not be used continuously in the stream of treated waters but only at appropriate intervals on a "shock dosing" and rinsing basis.

i) <u>Use of an oxidising agent</u>. The oxidant must have a good residual effect, which means that ozone is not suitable. A dose (0.5 ppm) of chlorine dioxide or potassium permanganate is appropriate.
ii) <u>Use of an algicidal agent</u>. The main reagent used is copper sulphate at doses of 0.1 to 0.2 mg/l (copper). At these levels, it is highly toxic to algae. At pH 7-8, the copper is easy to remove by means of flocculation/decantation/filtration. This method is mainly used in reservoirs.
iii) <u>Growth inhibiting factors</u>. Algae need light in order to grow, thus if sedimentation tanks are kept in the dark, algae cannot develop. Turbulent mixing techniques are currently used in reservoirs for this purpose. The removal of nutrients (precipitation of phosphorus) also constitutes a means to limit algal growth.

c) <u>Maintenance of filters</u>

Oxidants are used to restrict algae and bacterial proliferation on the surface of filtration granules. However, they do not necessarily have a beneficial effect on filtration, since this will depend on a number of other specific effects (the beneficial role of bacteria in biological filtration, for instance). Where necessary, instead of a continuous dosing of chemicals in treated waters, a shock dosing and rinsing at appropriate intervals (e.g. every six weeks) with an oxidant or a biocide (chlorine, chlorine dioxide etc.) is a useful process (28) for filter control, and it prevents organochlorine formation. The use of potassium permanganate and ozone for this purpose may, under certain conditions, cause side effects such as precipitation of manganese dioxide or excess gas release.

d) Destabilization of colloids by oxidising agents

Little is known about the destabilizing properties
of chlorine dioxide and permanganate. However, they are
likely to have a similar effect to chlorine (especially
chlorine dioxide) and tests are currently being carried
out. Most of the available data concerns the use of
ozone in pretreatment (29). The term "ozone micro-
flocculation" is often used for coagulation (or destabi-
lisation) by means of ozone. Ozone promotes flocculation
in many instances but the results obtained are often irregular. The higher the concentration of natureal organic
matter (humic substances) the clearer the effect of ozone.
If no organic matter is present, ozone has practically no
flocculating effect.

e) Removal of taste and odours

Ozone and chlorine dioxide, because of their higher
oxidising power, are more effective than chlorine in reducing taste and odours, and have been used especially
for this reason. Adsorptive treatments (activated carbon
for instance) are also particularly effective for a number of taste problems.

f) Colour removal

Here again both ozone and chlorine dioxide are more
effective than chlorine.

D. INTERACTION BETWEEN OXIDANTS AND ACTIVATED CARBON

As discussed earlier for activated carbon/chlorine
interaction, the relationship between ozone and activated
carbon granules is of interest. Like chlorine, ozone may
sometimes have a negative effect on adsorption (30); carbon sites are oxidised and become inactive; organic molecules become more polar and are less adsorbed.
However there is often a positive effect, on efficiency,
of activated carbon (30), (31). The main reason is the
dominant role of biological activity in the action mechanism of the carbon. Pre-ozonation makes organic molecules easier to break down by the micro-organisms active
in the carbon and this may be an important factor; the
lifetime of activated carbon is sometimes lengthened by
50 per cent. The term "biological activated carbon" is
sometimes applied.

Ozone does not appear to prevent bacterial growth,
even at the surface of the filter where the ozone concentration is high (32). Lower down in the filtration mass,
residual ozone disappears rapidly as it is broken down by
the carbon.

Ozone's stimulating effect on biological activity within the carbon bed seems useful, and combining the two treatments may constitute a promising development.

E. MANAGEMENT PROBLEMS IN THE USE OF OXIDISING AGENTS

It is well known that the transportation and handling of chlorine are difficult and dangerous, but on the other hand, it is easy to dose and to inject in the water. Most handling and transport problems are solved if chlorine is used in the form of hypochlorite. The only equipment needed is usually inexpensive and simply consists of a storage tank and a metering pump.

In general, the other oxidants are more difficult to use:

- Potassium permanganate must be dissolved which requires a large fixed tank and careful operation. It is difficult to measure permanganate accurately and any excess added to the treated water may have adverse consequences. For this reason, at least, permanganate is avoided unless manganese has to be removed from the water.
- Chlorine dioxide is also more difficult to apply than chlorine. In most cases, it is prepared before use from sodium chlorite and gaseous chlorine (*). The problems connected with gaseous chlorine storage therefore remain. Any reliable chlorine dioxide apparatus is relatively complex and includes several pumps and electro-valves. It is difficult to control the doses in relation to the flow of water to be treated.

_____

*) There are various methods for preparing chlorine dioxide but only one has a complete yield and does not leave untransformed reactants (chlorine, for instance) mixed with the chlorine dioxide. The recommended method for preparing chlorine dioxide is by reaction of sodium chlorite with a chlorine solution ($\geq$ 5 g/l). The efficiency of formation is then optimal (if the relative quantities of chlorine and chlorite are carefully controlled).

The other two methods of preparation, namely action of an acid either on sodium chlorite or on a mixture of sodium chlorite and sodium hypochlorite, only have a moderate efficiency of 80-90 per cent, and leave residual quantities of sodium chlorite or chlorine, which is unacceptable.

- Ozone is an oxidant which requires particular care from the point of view of plant design and construction. Ozonation facilities include:
  - an air drying plant with refrigerating equipment and a constantly regenerated dessicating gel;
  - a complex ozonation plant, (both mechanically and electrically);
  - an injection system (porous injector, turbine, etc.) in a contact structure where the water must remain for several minutes.

F. POSSIBLE NEGATIVE EFFECTS FROM THE USE OF OTHER OXIDANTS

Although relatively few data exist on this subject, it is likely that oxidants other than chlorine give rise to certain undesirable byproducts.

a) <u>Chlorine dioxide</u>

It has been suggested that under certain conditions organochlorines might be generated by the use of chlorine dioxide, but to a lesser extent than with chlorine. This may be the case in the presence of chlorine, which may thus negate the advantages of chlorine dioxide. Concern has also been expressed in relation to possible health risks from chlorite in drinking water, especially at pH levels above neutral, and the use of chlorine dioxide should be carefully planned in this respect. Further studies are needed to determine the level at which chlorite becomes a health risk. The health risks related to chlorite are not well known and might be comparable to those of nitrite; they might involve effects on certain blood constituents.

b) <u>Ozone</u>

It is known that oxidation byproducts may be produced by the action of ozone on organic substances, although in practice these are difficult to identify in drinking water because of their more polar nature. The possible health effects of such byproducts are not known.

c) <u>Chloramines</u>

Little is known about the possible byproducts of chloramines.

IV.8.   PRACTICAL APPROACHES FOR CONTROLLING
ORGANOHALOGEN COMPOUNDS

A. INTRODUCTION

Control of organohalogenated compounds can be achieved through various preventive and curative approaches which may be used separately or, preferably, associated.

a)   Preventive approaches

i) Where possible, a fundamental measure is to use raw waters of better quality which can enable the use of chlorine (or other oxidants) to be avoided or, at least, minimised.

ii) Alternative treatment processes (filtration, precipitation etc.) which avoid the use of chlorine (or other oxidants) at any stage are frequently possible. This approach is advisable for raw waters which are moderately or not polluted.

iii) A practical approach in a number of cases is to minimise chlorine application and to limit its use to final disinfection. This approach, which is appropriate for waters with a low precursor content, can be combined with the processes considered in (b). The use of other oxidants can also be considered.

iv) Reduction of the organic precursors as much as possible before chlorine application (at the very end of treatment) is a basic strategy in water treatment, especially where both the "precursor" content of raw waters and chlorine application may be substantial.

v) Careful control of both the conditions of raw water transportation and potable water distribution is often essential, as these may be major sources of organochlorines in drinking water, due to the addition of chlorine.

vi) Break-point chlorination should not be used, as it is a major producer of organochlorinated compounds.

b)   Curative approaches

Organochlorinated compounds, once formed, are generally very persistent and may pass through conventional
treatment without being satisfactorily removed.

i) Reduction of organohalogens already present in raw
waters. These substances generally only form a
minor part of pollutants in raw waters compared to
their potential levels in potable water; their levels in raw waters can be lowered by various processes which **are** discussed later.

ii) Reduction of organohalogens formed by chlorination
during treatment. As this is generally the major
source it is unrealistic and not sufficiently cost
effective to rely only on the removal of organochlorines at the final stage. However, as a "polishing" process, final removal of traces of organohalogens is a realistic and positive measure.
(The various methods available are discussed
later).

B.   DISCUSSION OF PREVENTIVE APPROACHES

a)   Selection of better raw waters and relocation of
water abstraction points as a practical means to reduce
the need for chlorination

With regard to the control of organochlorines (and
thus the final chemical and microbiological quality of
drinking water) this type of measure has several significant benefits:

- better quality raw waters will require less intensive applications of chemicals (chlorine) at all
  stages of transport, treatment and disinfection;
- they will generally contain less "precursors"
  (likely to react with the chemicals used);
- they will also contain less organohalogens
  initially.

Relocation of abstraction points presents a widespread problem for Member countries as, at present, many
potable water abstraction points are in urban and suburban areas (because towns have spread over the last few
decades). A necessary measure, which is frequently discussed, is to move the abstraction points upstream of urbanised zones; to be fully effective, however, this solution must satisfy several conditions:

- the relocation should be sufficiently far so that
  it does not have to be repeated after a few years;
- the abstraction point should not be downstream of
  sewage discharges, and continuous protection

against possible upstream pollution should be
ensured;
- chlorination should not take place while raw
waters are being transported. Where necessary, alternative processes, such as a primary clarification treatment and mechanical cleaning should be
used (see section VI. C.1).

b)   Alternative treatment processes not involving
chlorination

These processes were described in Chapter VI. They
consist mostly of biological and physical processes such
as biological filtration, adsorption, precipitation etc.,
and their combination. In some Member countries, the
strict application of these processes enables water to be
distributed without any application of chlorine (or other
oxidant).

c)   Conditions of chlorine utilisation for minimising
organohalogen formation

It is practical in a number of situations (in small
installations for instance) to minimise organochlorine
formation by a more controlled chlorine application.

i) this can be achieved by limiting chlorination to
its essential role i.e. final disinfection and
keeping the dose as low as possible;
ii) where good quality raw waters (with a low precursor content) are used, precautions are easier but
where the precursor content is rather high, precursor removal (see 4. below) should be practised.

d)   Removal of organic precursors

i) Aeration. This has little effect on the removal
of precursors.
ii) Storage (in open reservoirs). ROOK (3) has
studied the effectiveness of storage for three
weeks or more on the removal of organohalogen precursors. When carefully managed, these can be effective but nevertheless, it should not last too
long because the biomass may increase through eutrophication, as may the level of precursors.
The use of ground water and its compensed recharge
has shown to be effective in reducing the precursor level.
iii) Flocculation/separation. Depending on the reagent, and the separation technique used
(filtration/sedimentation), between 20 and 30 per
cent of the precursors can be removed.
iv) Adsorption.
- Powdered activated carbon. High doses are required for an appreciable effect and separation
of the carbon must follow.

- Granulated activated carbon. This is the most
  efficient treatment for the removal of precursors but its efficiency decreases with time
  /LOVE (33)7 because saturation soon occurs; thus
  regeneration must be carried out frequently.
- Macro-reticular resins have limited effects on
  humic acids which are the main haloform precursors. Basic reticular resins when fresh may
  remove up to 80 per cent of humic acids. Some
  problems are caused by the need for regeneration of these resins.
- Adsorption on aluminium oxides may also be used
  but has only limited efficiency.
  v) Oxidisation: A number of experiments /ROOK (3),
  LOVE (33)7 show that oxidants tend to increase
  the level of organohalogen precursors and thus
  cannot be used for their removal.
 vi) Biological filtration (on sand, carbon etc.).
  This process may remove substantial amounts of
  precursors. Its efficiency depends on: composition of the water, nature of the micro-pollutants
  present, pH, temperature, type of substrate, filtration velocity (or contact time) and filter
  depth. "Bank side" filtration (on natural soil)
  is used in several countries and is effective in
  reducing the level of precursors (28).

e)   Control of raw water transportation and potable
      water distribution

    As transportation of raw waters and distribution of
potable water are not considered as part of the treatment
itself, they often tend to be neglected as sources of organohalogen formation, although they may, in fact, be major contributors. For instance, high chlorine doses eqi-
valent to those of "break-point", may be used for raw
water transportation. Greater attention should be paid
to these underestimated sources of organochlorines, for
example:

- Chlorination of raw waters before transportation
  does not fulfil any essential role (it is aimed at
  controlling any biological growth and preventing
  deposits from fermenting); it should be replaced
  where necessary by alternatives, such as clarification of water before transportation, mechanical
  cleaning etc. (see Section VI. C.1).
- When and where a disinfectant residual (for bacteriostatic effect) is judged necessary in the potable water distribution network, it should be kept
  as low as possible, as organochlorine formation
  will continue in the pipes as long as precursors
  and chlorine are both present; less reactive oxidants (chloramines or chlorine dioxide) could also
  be used. Good maintenance and cleanliness of the

networks are of great importance; they will be a factor in microbiological safety and will enable the formation of organochlorines to be minimised (through lower dosing of chlorine residual and lower organic content in the pipes).

C.   DISCUSSION OF CURATIVE APPROACHES

Some "curative" processes are reviewed here for the removal of organohalogens present in raw waters, or those formed during potable water treatment. Their effectiveness is frequently low or irregular and depends on the types of organohalogens present (i.e. volatile or non-volatile).

a)   Biological treatment

Organohalogenated compounds are stable and very resistant to biodegradation. Thus biological treatment has little effect, but adsorption on the biomass may reduce their levels.

b)   Aeration

An air water ratio of about 20: (Love 33) is required at 20°C in order to remove about 80 per cent of the haloforms (i.e. 15 per cent of total oroganochlorines). ROOK (3).reports that cascades remove some of the chloroform in water but not all: The levels of non-volatile compounds (80 per cent) are probably unchanged.

c)   Storage

LOVE (33) and ROOK (3) report that storage for three weeks removes up to 40 per cent of haloforms in winter and 30 per cent in summer; again the non-volatile compounds are probably little affected.

d)   Flocculation/separation

Flocculation/separation treatment removes very few or no halomethanes.

e)   Adsorption

   i) Powdered activated carbon requires large doses of carbon and separation; this process is not cost effective.

ii) <u>Granulated activated carbon</u>. The results obtained
by LOVE (33) and the Paris Drinking Water Laboratory (34) - (see Tables 6 and 7) show that carbon
seems effective for the first three or four weeks
only, after which it is saturated and compounds
start to be leeched out. The use of activated
carbon is realistic only where organochlorines
are at very low levels (in the case of pesticides
for instance).

iii) <u>Adsorptive resins</u>. Macro-reticular resins may
be effective in the removal of haloforms (THMs).
Regeneration generally presents a problem except perhaps where steam can be used. However,
the specific use of these resins must be approved for water treatment producing drinking
waters.

The generalised use of these resins might present
the following problems: (1) they only remove the THMs
(the volatile organochlorines = 20 per cent of total);
(2) existing regulations are expressed for the moment
only in terms of THM; (3) thus the most important part
(the 80 per cent of non-volatile) may be overlooked and
remain uncontrolled. This example stresses the importance of expressing guidelines and standards in TOCl
(Total Organic Chlorine) and not in THM, which is an easy
but misleading parameter.

f)   <u>Oxidation of halomethanes</u>

None of the following oxidation treatments have been
shown to remove halomethanes efficiently:

Ozone, Chlorine dioxide, monochloramines, permanganate, UV rays, biological oxidation.

g)   <u>Reverse osmosis</u>

This method apparently has little or no effect.

D.   SUMMARY

The general strategies for organohalogen control can
be summarised as follows:

a) The primary objective should be to use raw waters
of the best available quality and to purify them
by methods which do not require chlorine or other
oxidants.

b) When the raw waters are of insufficient quality,
the use of chlorine, or other appropriate oxidants, should be considered for <u>final disinfection
only</u> (this should be their essential role); the
minimum dose compatible with microbiological
safety, should be used and the precursor content
should be minimised as much as possible
beforehand.

222

Table 6

REMOVAL OF PRECURSORS BY FLOCCULATION-DECANTATION

A. ALUMINIUM SALTS

$Al_2(SO_4)_3$ $18H_2O$ mg/l	Precursors $CHCl_3$ % of elimination	Precursors $CHBrCl_2$ % of elimination	WAC mg/l commercial solution	Precursors $CHCl_3$ % of elimination	Precursors $CHBrCl_2$ % of elimination
40	19	0	20	13	0
50	27	0	30	23	0
60	32	11	40	10	0
70	27	17	50	13	5
80	29	29	60	19	12

B. IRON SALTS

$FeCl_3$ mg/l Commercial solution	Precursors $CHCl_3$ % of elimination	Precursors $CHBrCl_2$ % of elimination	$FeClSO_4$ mg/l commercial solution	Precursors $CHCl_3$ % of elimination	Precursors $CHBrCl_2$ % of elimination
20	0	0	20	27	0
30	15	0	30	20	6
40	19	2	40	16	9
50	20	4	50	27	7.5
60	26	8	60	32	15.5

Table 7

HALOFORM REMOVAL BY FILTRATION ON GRANULATED
ACTIVATED CARBON

(5 minutes of contact time)

Time (in weeks)	$CHCl_3$	$CHCl_2Br$	$CHClBr_2$
1	100%	100%	100%
2	100%	100%	100%
3	100%	100%	100%
4	90%	100%	100%
5	75%	100%	100%
6	60%	100%	100%
7	50%	90%	100%
8	30%	80%	100%
9	0%	60%	100%
10	0%	50%	100%
11	-5%	30%	100%
12	-5%	25%	80%

IV.9.   COST OF CHLORINATION AND ITS ALTERNATIVES

A.   GENERAL

a)   Unit processes and overall treatment

Except in the case of water of very good quality,
which can be distributed without any treatment, normal
treatment consists of a series or sequence of additional
unit processes; these units are more or less interdependent and thus it is the overall process which should be
considered.

b)   Choice of combinations

From knowledge of the quality of raw water an optimum treatment process can be rationally designed. A
sequence which removes the pollutants present in raw
waters, but does not generate undesirable byproducts
(such as organochlorines) in treated water for public
supplies is essential.

For water of a given quality, a variety of combinations of unit treatment processes is generally possible.
Choice of the best sequence varies in every case, and can
only be achieved by a thorough study of the effectiveness
of each sequence in treating this specific type of water.

c)   Transformation of existing works

Most of the situations which water plant designers
will face in future will concern changes in existing
plants. Alternatives to, or the transformation of, installations will have to be approached in terms of the
additional cost related to the overall cost of the existing sequence.

d)   Treatment cost

It is difficult to give accurate figures for the
cost of each unit process since they all interact within
the overall process. Rough figures for the costs of unit
processes within treatment plants from several countries
can, however, be given.

B.   INSTANCES OF COSTS IN SOME TREATMENT PLANTS

Most examples given in Table 8 a-f (*) are for large
plants with rather sophisticated processes.

C.   GENERAL REMARKS ON THE ECONOMICS OF CHLORINE
REPLACEMENT

In the examples given above, which mostly concern
large plants, the replacement of chlorination by other
techniques would probably not appreciably change the overall treatment costs but it might increase the capital
costs.

In smaller plants, however, the introduction of carbon adsorption or ozone oxidation for instance, would
substantially raise both treatment and capital costs.
These processes would, therefore, result in a certain
cost for the water utilities. It must be remembered that
the cost of treatment is always small compared to the distribution cost; hence, even a considerable improvement in
the treatment process would only result in a marginal increase in the consumer's bill.

The reduction or abandonment of a chemical disinfection requires a certain reorientation of the distribution
strategy. On one hand a more careful approach would be
needed in the maintenance of the distribution network
(specially where repairs or leakages may occur), but on
the other hand, a more thorough treatment, better removal
of organics and a stricter clarification would keep the
whole system in better order and would mean less cleaning, less corrosion, less complaints from customers etc.

As a general rule, the distribution problems should
be included in the overall cost considerations because it
is unrealistic within a policy for improved drinking
water quality to disassociate the closely linked factors
i.e. raw waters, treatment and distribution.

Treatment costs are based in general on 1979 prices.
Unfortunately it is very likely that the use of the
treatment processes which are strongly energy dependent
(ozone, carbon reactivation) will probably increase rapidly in the near future. Whenever possible the use of
better quality raw waters will always be the safe policy
and the best alternative to the use of chlorination.

---

* In Table 8 a-f the data can, in general, be considered to be for 1978.

Table 8(a) Spain
Costs are given in 10^{-2} \$ USA/m^3

Treatment	Size of plant (in m^3 water/day)		
	50,000 m^3/day	5,000 to 50,000 m^3/day	5,000 m^3/day
Cl_2	1.45 (2.79)*	1.48	2.27
Cl_2 + GAC	(3.39)*		
Chloramines		1.97	2.80

* This plant treats polluted raw waters.
N.B. All these costs include operating and mainte-
nance costs but not the investment costs.
G.A.C. = granulated activated carbon.

Table 8(b) Switzerland
1979 Costs given in 10^{-2} Swiss Frs/m^3-
Surface water treatment

Treatment	Size of the Plant: 250,000 m^3 water/day Basis 40 million m^3/year		
	Capital	Operational	Total cost
Ozone 1.5 ppm	1.02	0.39	1.41
Chlorine dioxide 0.5 ppm	0.26	0.45	0.71
Chlorine	0.21	0.13	0.34
Filters	0.03		0.03
Reactivation			0.01
Total			0.04
Total water treatment			32.5

Table 8(c) Finland
Costs given in FMk/m³

Size of the Plant

Treatment	> 50,000 m³/day			5,000 to 50,000 m³/day			< 5,000 m³/day		
	Total cost	Operation	Capital	Total cost	Operation	Capital	Total cost	Operation	Capital
Cl₂		1.7(1)							
ClO₂		0.7(2)							
O₃	4	1.8	2.2	4.8(5) / 5.7	0.5(5) / 1.4	4.3 / 4.3			
GAC(6)	6.8(3)								
PAC(7)					1.5-6(4)				
Type of water	Polluted surface water								
Selling price of water	165			185					
Cost of distribution	87			110					
Cost of abstraction + treatment	70			104					
Administrative costs	20			13					

1) Dose 10 g/m³. 2) Dose 0.6 g/m³ NaClO₂. 3) Calculated costs. 4) 5 to 20 g/m³. 5) Ozone only used half of the year. 6) GAC = granulated activated carbon. 7) PAC = powdered activated carbon.

Table 8(d) France
Costs expressed in 10^{-2} FF/m³
Costs are given for low quality surface water treatment

Treatment	Size of Plant								
	> 50,000 m³/day			5,000 to 50,000 m³/day			< 5,000 m³/day		
	Total cost	Operational cost	Capital	Total cost	Operational cost	Capital	Total cost	Operational cost	Capital
Gas Cl₂ Disinfection 0.2 ppm									
Gas Cl₂ Prechlorination 6 ppm	1.6	1.4	0.20						
Na Hypochlorite 6 ppm	2.3								
ClO₂ 0.2 ppm				1.00	0.35	0.65	1.15	0.35	0.80
Ozone 0.5 ppm (Preoxidation)				2.00	0.8	1.2			
Ozone 2.5 ppm (Disinfection)	7.2	4.0	3.2						
Coagulation, Flocculation Alum. 80 ppm	7.0	5.0	2.0						
Filtration GAC(1) – Contractors Contact time 10 mm Reactivation/3 years	6.4	2.25	4.15						
Filtration on GAC – Contractors Contact time 10 mm 1 reactivation/year	10.3	6.7	3.6						
PAC(2) 20 ppm				6.50	5.5	1.0			
Biological Nitrification				6.5	2.50	4.0			
Average selling price of water	250								

1) GAC = granulated activated carbon.
2) PAC = powdered activated carbon.

Table 8(e) Netherlands
Costs are given in Dutch Cents/m³ -
Surface water treatment

Treatment	Size of the Plant: m³/day				
	12,000	29,000	55,000	120,000	300,000
CHLORINATION					
Capital	(non-existant)	1.28			0.33
Labour		0.88			0.011
Liquid chlorine 15 ppm		0.74			
Chlorine at 1 ppm		0.05			0.05
Total		2.90			0.74
Distribution Costs		49			56
OZONISATION					
Capital		1.67		1.62	
Energy		0.95		0.90	
Labour		0.43		0.40	
Total		3.05		2.92	
Distribution Costs		49-50		56	
GRANULATED ACTIVATED CARBON					
Apparent contact time	7 min.		40 min. 20 min.	12 min.	
Capital: (Amortisation – 15 years, Equipment – 15 years, Engineering – 40 years)	3.83		7.78 4.29	2.17	
Operational	0.5		0.9 0.9	0.62	
Regeneration	0.97 (2-3 mths.)		3.6 1.8 (12 mths.)	0.77 (12 mths.)	
Total	5.3		12.3 8.0	3.56	

Table 8(f) United Kingdom
Costs are given in pence/m³ - Predictions

Treatment	Size of the Plant		
	10,000 m³/day	100,000 m³/day	300,000 m³/day
Granulated activated Carbon (GAC) (dose 60 g/m³)			
Capital costs £/100 m³/day	70,000	29,000	21,600
Operational costs and maintenance pence/m³	2.2	1.1	1.0
Total cost for GAC (reactivation on site) 10% per year on {equipment 15 years / civil engineering 30 years} pence/m³	4.0	2.0	1.5
Ozone Dose 3 ppm in pence/m³ Contact time 5 minutes 10% per year on {equipment 15 years / civil engineering 30 years} Energy use: 2 pence/Kwh, 25 Kwh/kg ozone	0.8	0.5	0.4
Chlorine Dose: 0.5 ppm			
Chemical cost:	0.01	0.007	0.005
Total cost:	0.05	0.05	0.05
Chlorine dioxide Dose 0.5 ppm pence/m³			
Chemical cost: pence/m³	0.11	0.11	0.11
	70,000	29,000	21,600

Note: If GAC treatment is used, the increase in the consumer's water bill would be very modest (2 per cent). However, the impact on capital expenditure would be more significant.

D. EXAMPLES OF TREATMENT PRACTICES AND THEIR COST
    WHEN ALTERNATIVES TO CHLORINATION ARE USED

a) Water of good chemical quality

    i) If there is no danger of bacterial contamination,
       even during distribution, disinfection may be
       abandoned. This means, however, that stringent
       disinfection precautions must be taken when ex-
       tending or maintaining the system.
   ii) Where there is danger of contamination, disinfec-
       tion is desirable:
       - if no residual effect is desired (where there
         is bacterial contamination at the source but
         not in the network), ozone and UV irradiation
         are two possible methods of treatment.
       - if instead a residual effect is needed because
         of potential contamination during distribution,
         chlorine dioxide or chloramine can be used.
       Average doses for disinfection of drinking water
       of good quality are about 0.2 ppm chlorine, 0.1
       0.1 ppm ClO_2 or 0.5 ppm ozone. A change from
       chlorine to chlorine dioxide would incur an ad-
       ditional treatment cost of 0.24 U.S. cents/m^3.
       The use of ozone would bring a cost increase over·
       chlorination of 0.40 U.S. cents/m^3.

       The investment cost is proportionately higher in
       the case of ozone whereas the operational cost is
       higher for ClO_2.
  iii) If water contains ammonia and there is also a
       danger of bacterial contamination, the following
       treatment practices may be envisaged:
       - biological nitrification followed by a light
         chlorine dioxide or chlorine application /see
         case (a)7;
       - exceptionally, if the content of "precursors"
         is very small or during cold periods, break-
         point chlorination (chlorine dose up to ten
         times the ammonia weight) may be envisaged. It
         should then be carried out at the end of treat-
         ment, providing final disinfection at the same
         time.
       Costs will, of course, depend on the amount of
       ammonia to be eliminated.

    For water with low levels of ammonia, the replace-
ment of break-point chlorination by biological nitrification within specific treatment works would thus multiply
tenfold the treatment cost. One must bear in mind, however, that biological nitrification can be partly carried
out in the conventional water treatment works (settling
tanks and rapid gravity filters) and in storage basins in
a pretreatment stage.

232

	Costs in US$ per m^3	
	Capital	Operating cost per gramme of NH_4 eliminated
Biological nitrification	0.01	0.0005
Break-point chlorination	0.001	0.0015

The water may also contain iron and manganese - requ ring complementary oxidising treatment - and sometimes taste-causing substances calling for activated carbon treatment. The removal of ammonia can then be combined with the removal of the other substances (e.g. rapid biological filters for Fe-M -NH_4 removal).

b) **Moderately polluted surface water containing precursors (e.g. humic acids)**

The following treatment sequences may be envisaged:

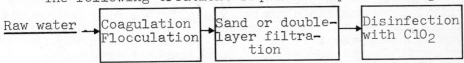

Raw water → Coagulation Flocculation → Sand or double-layer filtration → Disinfection with ClO_2

The cost of coagulation-flocculation treatment (at 10 ppm) + filtration is about US$ 0.005 per m^3 (operating cost) and US$ 0.006 per m^3 (capital cost).

The total cost price of treated water is therefore approximately US$ 0.012 per, m^3.

The introduction of the coagulation-flocculation-filtration stage results in some increase in the treatment cost. This is a good way, however, to produce potable water containing low levels of organochlorinated compounds from a moderately polluted surface water.

Direct filtration may not be sufficiently efficient for a good removal of organics; when their concentration increases it is preferable to use sedimentation (or flotation) with larger coagulant doses prior to filtration.

c) **Raw waters of low quality (polluted surface waters)**

The treatment processes used for such waters will be similar to those used for moderately polluted waters, but will be more intensive. The combinations of treatment will be more sophisticated, and will include advanced clarification, intermediate adsorption or oxidation etc.

The cost will, of course, be somewhat higher, especially if activated carbon or ozone are extensively used. Replacement of chlorination, especially when used as a pretreatment (break-point chlorination) does not result, in most cases, in an appreciable cost increase.

E.   CONCLUDING REMARKS

The cost of water treatment is generally a small fraction of the cost of drinking water to consumers. In many cases minor modifications to existing treatment aimed at minimising the precursors before the application of oxidants, without endangering the biological quality of the water, would be effective, for little or no cost, in substantially reducing the byproducts formed. As the cost involved is usually moderate, it is prudent to carry out these modifications where feasible.

Using oxidants other than chlorine may result in somewhat increased water treatment costs although still only a very small and probably negligible increase in the total cost of the water supplied. The use of chloramines would not appreciably alter costs; ozone would require somewhat increased capital expenditures (compared to chlorine); whereas chlorine dioxide would require low initial capital but incur higher operation costs.

The use of certain treatments, such as granular activated carbon or resins to remove organochlorine by products after their formation, would be by far the most costly option, and probably only needs to be considered as a "polishing" technique when water quality is so poor that other conventional technologies cannot sufficiently reduce the oxidant demand for precursors.

Control options available for small water systems may be considerably different to those for large systems because small systems have higher per capita costs, less access to trained operating personnel and less capacity to perform sufficient operational monitoring. The use of high quality raw waters is thus particularly important in this case as it makes the entire treatment and distribution process easier and safer.

IV.10. CONCLUSIONS

Although organochlorinated substances may already be present as pollutants in raw waters, chlorination of water containing natural or synthetic organic precursors is the main source of halogenated organic chemicals in drinking water, as most surface waters and some ground waters contain substantial amounts of precursors.

Trihalomethanes (including chloroform) are currently the more easily identified organohalogen by-products, but normally they only represent a modest proportion (approximately 20 per cent) of total organohalogens present in drinking water, and not necessarily the most hazardous substances. Although a large proportion of organohalogens present in drinking water are still practically unidentified, useful overall or partial parameters have been developed and these permit a closer assessment of the presence or potential formation of organohalogens in drinking water. Total Organic Chlorine (TOCL) is the most relevant test as it is comprehensive and applies to the whole range of organochlorines present (however, it does not individually identify compounds). Total Organic Carbon (TOC) is a useful complementary test for assessing the potential amounts of "precursors". The Trihalomethane (THM) analysis is a relatively easy test but only gives a partial view of the total mix of chemicals present.

Oxidants such as chlorine, ozone, chlorine dioxide and to a lesser extent chloramines, are effective reagents in drinking water treatment, especially for disinfection which is their essential function. However, being chemically very active, they may produce a variety of by-products by reacting with the organic "precursors" present in waters. Up to now, organochlorinated by-products have received most of the attention for a number of reasons: they are frequently encountered in significant levels in drinking water; a number of organochlorines are known or suspected to present health hazards; and they can currently be detected with available techniques. Although knowledge is very limited, it would also be prudent to consider the possible effects of the by-products which may arise from the use of other oxidants.

In many drinking water treatment installations, chlorine is used extensively throughout the system from the initial raw water transportation to final drinking water distribution. It is clear that chlorine applications, particularly from the early stages, when water may still contain substantial levels of organic precursors, will lead to significant organochlorine formation. A better control of organohalogens in drinking water requires more prudence and selectivity in the use of chlorine, which should, as far as possible, be kept to its essential role of final disinfection. For the bacteriostatic effect in the network, a residual of chlorine dioxide or chloramine is an effective alternative.

In principle, processes which remove or reduce contaminants such as physical and biological treatment should be preferred to processes such as chemical treatment which transform them into other chemicals with undesirable or unknown effects. Member countries should also specify and control the quality of additive chemicals used in potable water treatment.

The gradual decrease, frequently noted in the quality of raw waters used over the past few decades, has resulted in the intensification of treatment. The parallel increase of both organic pollutants in waters and chlorine applications all along the treatment system has lead to the organohalogen levels currently encountered in drinking waters. The use of good quality raw waters is thus a fundamental policy for the control of organochalogens and the other trace pollutants in potable water.

Break-point chlorination, which is commonly practised for ammonia removal, may lead to high levels of organochlorines in drinking water. Thus it seems advisable to use other ammonia removal methods such as biological removal, storage, resins, better protection of the source or combinations of these processes. Under exceptional circumstances (e.g. during cold periods) when break-point chlorination is used, it should be carried out as a final treatment stage, after removal of organic precursors, providing final disinfection at the same time.

Under certain geographical and geological conditions, raw waters of good quality may, however, have a high content of humic and fulvic acids. Although these substances may not in themselves present a real hazard to human health, they react readily with chlorine to form organochlorinated compounds. Precautions should be taken with these waters so that the processes used throughout the water transportation, treatment and distribution system, minimise the formation of organohalogens. Similar caution is required with sources subject to sea water intrusion and bromide contamination, as chlorination will lead to the formation of significant amounts of both organobromines and organochlorines.

Where chlorine is used for purposes other than final disinfection (e.g. keeping the treatment plant free of biological growth) alternative approaches should be adopted (such as "shock-dosing" and rinsing of installations at appropriate intervals).

Chlorination of raw waters during transportation, carried out for secondary purposes only (control of fixed organisms) may be a very important source of organohalogens which is generally underestimated or neglected because it does not take place in the plant. A number of alternative processes such as clarification of water before transportation, mechanical cleaning, shock dosing and rinsing, etc., can be used successfully.

When a distribution system is in poor condition, high chlorine dosing is often used to maintain a substantial disinfectant residual, and in the presence of precursors the formation of organohalogens will continue as long as chlorine persists in the system. Good maintenance and cleanliness of the distribution network is a contribution to biological and chemical safety of water as it permits to minimise or avoid the use of a chlorine residual.

The microbiological quality of drinking water is, like its chemical quality, of prime importance, and biological safety should not be compromised when improving the chemical quality. Sufficient technologies are available to optimise both biological and chemical purity of water at a cost which, specially for larger water systems, is generally a modest fraction of the consumer's cost for drinking water. Thus it is false economy to sacrifice drinking water quality by not applying optimal treatment.

The increased risk of cancer due to organochlorinated compounds in drinking water cannot yet be fully evaluated. Besides the estimates on health risks from chloroform, knowledge is still lacking on the potential hazards from the large number of unidentified organohalogenated compounds encountered in drinking water. As there may be no "safe" level for these substances, prudence is required and it is justifiable to maintain their concentration at the lowest possible level.

Member countries should encourage measures to control drinking water contamination by chlorinated products (as well as other by-products which may be generated by chemicals, materials or practices used in drinking water treatment and supply). These measures could include standards, guidelines and recommendations for management practices and technologies in water systems, along with emphasis on the quality of raw water sources. The objectives should include minimisation of by-products to the extent feasible while maintaining bacteriological safety and taking into account the size and type of plant, geographical conditions, etc. Unless the goals reflected in guidelines are high enough they may have a negative effect for a large number of water works already within the limit by acting as a disincentive for any further improvement. Periodic updating should be foreseen as additional toxicological data become available.

IV.11. PROPOSED SUBJECTS FOR RESEARCH

Identification of organohalogens other than the THM (Trihalomethane); and assessment of their hazard to human health.

Aggregate toxicity of TOCl (Total Organic Chlorine) products in drinking water.

Development of methods for automated analysis and detection of TOCl; TOC; THM.

Methods for removal of organochlorines from water: 1) already present in raw waters; 2) and those produced by treatment.

Identification of organic by-products (other than organochlorines) arising from the application of oxidants, and their health effects.

Investigation of the conditions of formation of chlorates and chlorites from chlorine dioxide application and their effects on human health.

TOCl formed by chlorine dioxide and chloramine applications.

Toxicology of brominated by-products in potable water.

Disinfection techniques for drinking water not involving chemical oxidants.

Special attention should be given to development of treatment and monitoring technologies appropriate for use by small drinking water systems.

Optimisation of methods for: slow sand filtration, bank filtration and recharge of aquifers (with clean water) eliminating the use of chlorine.

Influence of the use of eutrophied waters on the presence of precursors and the formation of organohalogens; influence of pH of water on volatile or non-volatile organohalogen formation.

239

Annex

IV.1 DEFINITIONS

Precursors are natural or synthetic organic compounds capable of reacting with chlorine (and other halogens) and producing organochlorinated compounds (or more generally organohalogenated compounds).

Organohalogens (or organohalogenated compounds): organic compounds whose molecule contains one or more halogens (such as chlorine, bromine and iodine).

    i) Volatile organohalogens: the molecule contains less than four atoms of carbon.

    ii) Non-volatile organohalogens: the molecule contains four or more carbon atoms.

Organochlorines (or organochlorinated compounds): organic compounds whose molecule contains one or more chlorine atoms.

Trihalomethanes (THMs) are volatile organohalogens whose molecule contains one atom of carbon, one atom of hydrogen and three atoms of halogen.

    i) When there are three atoms of chlorine it is Chloroform

    ii) When there are three atoms of bromine it is Bromoform

    iii) For instance when there is one atom of bromine and two atoms of chlorine, it is Monobromodichloromethane, etc.

Annex

IV.2 BIBLIOGRAPHY

1. ROOK J.J., 1974. Formation of Haloforms during
    Chlorination of Natural Waters.
    Water Treatment Exam, 1974, 23, 234-243.

2. BELLAR T.A., LICHTENBERG J.J., KRONER R.C. The
    Occurrence of Organohalides in Chlorinated Drinking
    Waters.

3. ROOK J.J., 1976. Haloform in Drinking Water
    JAWWA 1976, 68, No. 3, 168-172.

4. DONALDSON W., 1922. Observation of Chlorination
    Tastes and Odours. Engineering and Contracting
    Water Works - Monthly Issue May 1922, 74-78.

5. BUNKER G.C., 1929. The Use of Chlorine in Water
    Purification. J.Am. Assn., 1929, 92, 1-18.

6. ENSLOW L.H., 1934. Modern Water Chlorination
    Practice. J.N.E. Water Works Assn. 1934, 48, 6, 22.

7. FALES A.S., 1926. Progress Report on Recent
    Developments in the Field of Industrial Wastes
    in relation to Water Supply. Committee 76 AWWA
    JAWWA 1926, 16, 302/329.

8. STREGTER H.W., 1929. Chlorophenol Tastes and
    Odours in Water Supplies of Ohio River Cities.
    Public Health Reports 1929, 44, 2149-2156.

9. ROOK T.J. et al. Von Wasser, rr. 1975, p. 23-30
    and 57-63.

10. BUSH B., NARANG R.S., SYROTYNSKI S. 1977. Screen-
    ing for Halo-organics in New York State Drinking
    Water.
    Bull - Environ. Contom. Toxicol. 18 No. 4 - 436.

11. KÜNH W. and SANDER R., 1978. Vorkommer und Bestim-
    mung Leichflüchtiger Chlorkohlenwasserstoffe.

12. BOHN B., GAJEK H., SONNEBORN M., 1978. Flüchtige
    Halogenkohlenwasserstoffe in Verschiedenendeutschen
    Trinkwassern.

13. MONTIEL A., 1978. Les halométhanes dans l'eau - Mécanismes de formation - Evolution - Elimination. These de Doctorat d'Etat présentée le 23 mai 1978 à l'Université Pierre et Marie Curie, Paris VI.

14. PIET G.J., ZOETEMAN B.C.J., SLINGERLAND P., 1978. Der Effekt der Chloung auf die Trinkwasserqualität.

15. MAES M., 1975. Livre: Les résidus industriels - Traitement - Valorisation. Législation. Tome I. Technique et documentation, 11 rue Lavoisir, Paris 8ème.

16. ANONYME 1973. Akzo-Zout-Chemie-Nederland-B.V. Chlorinated Hydrocarbons - The Market in Europe. EU - Chem. - News - Chemiscope - Supp. 24th October 12-67-70 1973.

17. ANONYME 1972. Market - Oxygenated Solvents Forecasts show modest growth. Chem. age - Land - 104 (May)18 - 1972.

18. ROE F.J.C. cited by TARDIFF R.G. J.Am. Water Works Assoc. 69, 651 (1977).

19. N.C.I. Carcinogenesis Bioassay of Chloroform. Nation Institute, Bethesda, M.D. 1976.

20. DEBIAS M. 1974. Ministère de la Qualité de la Vie cité par Maes M., Les Résidus Industriels - Traitements Valorisation. Législation Tome I - Technique et Documentation II, rue Lavoisier, Paris 8ème 1975.

21. LOVELOCK J.E., MAGGS R.J., WADE R.J., 1973. Halogenated Hydrocarbons in and over the Atlantic Nature 241-194-196, 1973.

22. WHITE, G.C. Handbook of Chlorination. Von Nostrand Reinhold, New York, New York, 1972.

23. STREETER H.W. and C.T. WRIGHT. Prechlorination in relation to the efficiency of water filtration processes. J. AWWA 23, 22, 1931.

24. WESTON R.S. 1924. The Use of Chlorine to assist Coagulation. JAAWA 11, 446.

25. FIESSINGER F., TAMBO N. La Coagulation et la Floculation. 12 Congrès de l'Association Internationale des Distributeurs d'Eau, Kyoto. October 1978.

26. SONTHEIMER H. et al. The Mulheim Process. JAWWA 393, July 1978.

27. MACLAREN J.R., FARQUHAR G.I. Factors affecting ammonia removal by clinoptilolite. J. Environ. Eng. Div. 429, 446, 1973.

28. ALTMANN H.J., GROHMANN A., HASSELBARTH U., KOWALSKI H. SARFERT F. Grundwasseranreicherung für die Trinkwassergewinnung mit der Verfahrenskombination Flockung-Filterung-Bodenpassage (Versuchsanlage Jungfernheide).
Teil I: Untersuchungsergebnisse und deren Bedeutung für die Trinkwasserversorgung Berlins
Teil II: Bewertung der Reinigungsleistung
Internationales Symposium Künstliche Grundwasseranreicherung, Dortmund, 14th - 18th May, 1979 (im Druck).

29. FIESSINGER F.J., MALLEVIALLE and ALLIZADEH A. Flocculant effect of Ozone. IOA Conference, Nice, January 1979.

30. RICHARD Y. and FIESSINGER F.J. Complementary Role of Ozone and Active Carbon, IOI Conference, Paris June 1977.

31. GUIRGUIS W.A., PROBER R., SLOUGH J.W. Effects of Ozone Pre-treatment and Bacterial Growth on Activated Carbon Treatment. ACS Symposium on Activated Carbon, Miami Beach, Florida, September 1978.

32. BENEDEK A. The Effect of Ozone on Activated Carbon Absorption. A Mechanistic Analysis of Water Treatment Data. IOI Symposium, Toronto, November 1977.

33. LOVE O.T., CARSWELL J.K., MILTMER R.J., SYMONS J.M., 1976. Treatment for the Prevention or Removal of Trihalomethanes in Drinking Water, Appendix - Interim Treatment Guide for the Control of $CHCl_3$ and other Trihalomethanes Water Supply Research Division, Cincinnati, 1976.

34. Paris Drinking Water Laboratory. MONTIEL A. Les Goûts de l'Eau. Conférence presentée le 22nd September, 1977 au Journée Internationales de Pharmacie. Paris 19th - 23rd September, 1977.

35. TRUHAUT R., GAK J.C., GRAILLOT Cl. Recherches sur les risques pouvant resulter de la pollution chimique des eaux d'alimentation. Water Research, Vol. 13 No. 8 1979, p. 689.

36. LOVELOCK J.E., 1974. Atmospheric Halocarbons and Stratospheric Ozone. Nature 252-292-294, 1974.

37. VILAGINES R., MONTIEL A., DERREUMAIX A., LAMBERT M., 1977. Etude Comparative de la Formation des Halo- méthanes lors du Traitement de l'Eau Potable par le Cjlore et ses Dérivés dans les Usines de Traite- ment d'Eau Potable et d'Eaux Usées. 96ème Annual A.W.W.A. Conférence Anaheim, 8 mai 1977, Désinfection Séminar.

38. E.P.A. 1975. Preliminary Assessment of Suspected Carcinogen in Drinking Water. Rpt. to United States Congress, EPA, Washington D.C., December 1975.

39. SONNEBORN M. (Edit.), Gesundheitliche Probleme der Wasserchlorung und Bewertung der dabei gebitdeten halogenierten organischen verbindungen. Wabolu Bericht 3/1979, Dietrich Reimer Verlag, Berlin 19.

40. SONNEBORN M., and BOHN B., 1977. Formation and Occurrence of Haloforms in Drinking Water in the Federal Republic of Germany. R. JOLLEY (Edit.), Water Chlorination, Environmental Impact and Health Effects. Vol. 2. Ann Arbor Science 1978.

41. ANONYME 1972. Solvent growth rates are crambled by ecology. Chem. Marketing Reported - 4th - 7th May, 1972.

42. SYMONS J.M., 1975. National Organics Reconnaissance Survey for Halogenated Organics. JAWWA 1975, 67, 11, 634.

43. LIETZKE, M.H., 1978. Umweltbee influssung und gesundheitliche auswirküng der chlorung des wassers aus amerikanischer sicht. Wabolu Berichte 3, 13-20.

44. MORRIS J.C., 1975. Formation of Haloginated Organics by Chlorination of Water Supplies. EPA 600/1-75-002 United States EPA, Washington D.C. 1975.

45. BUNN W.W., HAAS, B.B., DEAME E.R., KLEOPPER R.D., 1975. Formation of Trihalomethanes by Chlorination of Surface Water. Environmental Letters, 1975, 10, 3.

46. ROOK J.J., 1975. Bromierung Organischer Wasserin- haltstoffe als Mehen-reaktion der Chlorung Vom Wasser. 1975, No. 44, 57-67.

47. PATIN A.T., 1945. The Determination of Free Chlorine and of Chloramine in Water using p- Arrinodimethylaniline. Analyst England 1945, 70, 203.

48. SAUNIER B., SELLECK R.E. 1976. Kinetics of Break-
     point Chlorination and of Disinfection. Report
     No. 76, 2nd May, 1976. Sanitary Engineering Re-
     search Laboratory. College of Engineering and
     School of Public Health University of California,
     Berkley.

49. COLEMAN W.E., LINGG R.D., MELTON R.G., KOPFLER F.G.
     The Occurrence of Volatile Organics in Five Drink-
     ing Water Supplies. United States EPA Health
     Effects Research Laboratory Water Quality Division.
     Cincinnati, Ohio 45268.

50. MORRIS J.C. 1975. Formation of Haloginated Organics
     by Chlorination of Water Supplies. A review NTIS
     United States Department of Commerce 1975,
     No. PB 241, 511.

51. HARRISON R.M., PERRY R., WELLINGS R.A., 1976. Ef-
     fect of Water Chlorination upon Levels of some Poly-
     nuclear Aromatic Hydrocarbons in Water. Environ.
     Sci. Technol. 1976, 10, No. 12, 1151-1156.

52. STEVENS A.A., SEEGER D.R., SLOCUM C.J., 1976.
     Products of Chlorine Dioxide Treatment of Organic
     Materials in Water. Presented at the Congress on
     Ozone/Chlorine Dioxide Oxidation Products of
     Organic Materials, 17th - 19th November, 1976,
     Cincinnati, Ohio.

53. MILTNER R.J., 1977. Measurement of Chlorine Dioxide
     and Related Products. Water Supply Research Divi-
     sion, Municipal Environmental Research Laboratory
     United States EPA, Cincinnati, Ohio, 1977.

54. MUSIL J., KNOTECK Z., CHALUPA J., SCHMIDT P., 1964.
     Toxicological Aspect of Chlorine Dioxide Application
     for the Treatment of Water containing Phenol.
     Scientific Papers from Inst. Chem. Tech., Prague
     1964, 8, 327.

55. HEFFERMAN W., cité par MILTNER R.J., 1977. Measure-
     ment of Chlorine Dioxide and Related Products.
     United States EPA, 1977.

56. SIMMON, V.F., TARDIFF, R.G., 1978. The Mutagenic
     Activity of Halogenated Compounds found in Chlorin-
     ated Impact and Health Effects.
     Editor: Ann Arbor Science.

57. ALAVANJA M., GOLDSTEIN I., SUSSER M., 1978.
     A Case Control Study of Gastrointestinal and Urinary
     Tract Cancer Study: Mortality and Drinking Water
     Chlorination. Water Chlorination - Environmental
     Impact and Health Effects. Editor: Ann Arbor
     Science.

58. RILEY T.L., NANCY K.H., BOETTNER E.A., 1978.
The Effect of Pre-ozonation on Chloroform Production
in the Chlorine Disinfection Process.
Water Chlorination - Environmental Impact and
Health Effects. Vol. 2. Editor: Ann Arbor Science.

59. DRAPEAU A.J., TRUDEAU M., 1975. Du Chloroforme
dans votre Eau Potable. Eau du Québec 1975.
Vol. 8, No. 2.

61. EPA. Analytical Report New Orleans Area-Water
Supply. Study EPA 906/10.74002. Surveillance and
Analysis Division United States EPA Region VI Dallas,
Texas.

62. DUNHAN L.J., O'GARA R.W., TAYLOR F.B., 1967. Studies
on Pollutants from Processed Water Collection from
three Stations and Biologic Testing for Toxicity
and Carcinogenesis, Am. J. Pub Health 1967, 57 (12),
2178-2185.

63. DORN M.F., CUTLER S.J., 1959. Morbidity from Cancer
in the United States. PHS. MONOGR. No. 56,
Washington D.D. Gov. Ptg. Office, 1959.

64. CONNEL G.F., Mac, ERGUSON D.M., PEARSON G.R., 1975.
Chlorinated Hydrocarbons and the Environment.
Endeavour January 1975, Vol. 34.

65. EPA. Interim Primary Drinking Water Regulation Proposed Regulation for Chloroform and other Trihalomethanes in Drinking Water.

66. United States DHEW PHS. CDC. NIOSH., 1974. Criteria
for a Recommended Standard Occupational Exposure to
Chloroform.

67. TOUSSAINT, 1972. Le Bioxyde de chlore et le Traitement des Eaux. Tribune du Cebedeau 1972, No. 342,
263-264.

68. WIDEMANN O., 1957. 4 Jahre Praktische Erfahrung
mit Chlordioxyd. Vom Wasser. XXIV Band 1957,
50, 70.

69. FROEHLER H., 1961. Production of Chlorine Dioxide
from Sodium Chlorite. Wasser. Luft. Betrieb.
1961, 5, 332-336, 380-387.

70. MONTIEL A., MOUCHET J., 1979. Utilisation de
l'Ozone en Stérilisation. Eau et Industrie.

71. EVANS, F.L., 1972. Publication: Ozone in Water
and Wastewater Treatment. Edition: Ann Arbor
Science.

OECD SALES AGENTS
DÉPOSITAIRES DES PUBLICATIONS DE L'OCDE

ARGENTINA – ARGENTINE
Carlos Hirsch S.R.L., Florida 165, 4º Piso (Galería Guemes)
1333 BUENOS AIRES, Tel. 33.1787.2391 y 30.7122
AUSTRALIA – AUSTRALIE
Australia and New Zealand Book Company Pty, Ltd.,
10 Aquatic Drive, Frenchs Forest, N.S.W. 2086
P.O. Box 459, BROOKVALE, N.S.W. 2100
AUSTRIA – AUTRICHE
OECD Publications and Information Center
4 Simrockstrasse 5300 BONN. Tel. (0228) 21.60.45
Local Agent/Agent local :
Gerold and Co., Graben 31, WIEN 1. Tel. 52.22.35
BELGIUM – BELGIQUE
CCLS – LCLS
19, rue Plantin, 1070 BRUXELLES. Tel. 02.512.89.74
BRAZIL – BRÉSIL
Mestre Jou S.A., Rua Guaipa 518,
Caixa Postal 24090, 05089 SAO PAULO 10. Tel. 261.1920
Rua Senador Dantas 19 s/205-6, RIO DE JANEIRO GB.
Tel. 232.07.32
CANADA
Renouf Publishing Company Limited,
2182 St. Catherine Street West,
MONTRÉAL, Que. H3H 1M7. Tel. (514)937.3519
OTTAWA, Ont. K1P 5A6, 61 Sparks Street
DENMARK – DANEMARK
Munksgaard Export and Subscription Service
35, Nørre Søgade
DK 1370 KØBENHAVN K. Tel. +45.1.12.85.70
FINLAND – FINLANDE
Akateeminen Kirjakauppa
Keskuskatu 1, 00100 HELSINKI 10. Tel. 65.11.22
FRANCE
Bureau des Publications de l'OCDE,
2 rue André-Pascal, 75775 PARIS CEDEX 16. Tel. (1) 524.81.67
Principal correspondant :
13602 AIX-EN-PROVENCE : Librairie de l'Université.
Tel. 26.18.08
GERMANY – ALLEMAGNE
OECD Publications and Information Center
4 Simrockstrasse 5300 BONN Tel. (0228) 21.60.45
GREECE – GRÈCE
Librairie Kauffmann, 28 rue du Stade,
ATHÈNES 132. Tel. 322.21.60
HONG-KONG
Government Information Services,
Publications/Sales Section, Baskerville House,
2/F., 22 Ice House Street
ICELAND – ISLANDE
Snaebjörn Jönsson and Co., h.f.,
Hafnarstraeti 4 and 9, P.O.B. 1131, REYKJAVIK.
Tel. 13133/14281/11936
INDIA – INDE
Oxford Book and Stationery Co. :
NEW DELHI-1, Scindia House. Tel. 45896
CALCUTTA 700016, 17 Park Street. Tel. 240832
INDONESIA – INDONÉSIE
PDIN-LIPI, P.O. Box 3065/JKT., JAKARTA, Tel. 583467
IRELAND – IRLANDE
TDC Publishers – Library Suppliers
12 North Frederick Street, DUBLIN 1 Tel. 744835-749677
ITALY – ITALIE
Libreria Commissionaria Sansoni :
Via Lamarmora 45, 50121 FIRENZE. Tel. 579751/584468
Via Bartolini 29, 20155 MILANO. Tel. 365083
Sub-depositari :
Ugo Tassi
Via A. Farnese 28, 00192 ROMA. Tel. 310590
Editrice e Libreria Herder,
Piazza Montecitorio 120, 00186 ROMA. Tel. 6794628
Costantino Ercolano, Via Generale Orsini 46, 80132 NAPOLI. Tel.
405210
Libreria Hoepli, Via Hoepli 5, 20121 MILANO. Tel. 865446
Libreria Scientifica, Dott. Lucio de Biasio "Aeiou"
Via Meravigli 16, 20123 MILANO Tel. 807679
Libreria Zanichelli
Piazza Galvani 1/A, 40124 Bologna Tel. 237389
Libreria Lattes, Via Garibaldi 3, 10122 TORINO. Tel. 519274
La diffusione delle edizioni OCSE è inoltre assicurata dalle migliori
librerie nelle città più importanti.
JAPAN – JAPON
OECD Publications and Information Center,
Landic Akasaka Bldg., 2-3-4 Akasaka,
Minato-ku, TOKYO 107 Tel. 586.2016
KOREA – CORÉE
Pan Korea Book Corporation,
P.O. Box nº 101 Kwangwhamun, SÉOUL. Tel. 72.7369

LEBANON – LIBAN
Documenta Scientifica/Redico,
Edison Building, Bliss Street, P.O. Box 5641, BEIRUT.
Tel. 354429 – 344425
MALAYSIA – MALAISIE
and/et SINGAPORE - SINGAPOUR
University of Malaya Co-operative Bookshop Ltd.
P.O. Box 1127, Jalan Pantai Baru
KUALA LUMPUR. Tel. 51425, 54058, 54361
THE NETHERLANDS – PAYS-BAS
Staatsuitgeverij
Verzendboekhandel Chr. Plantijnstraat 1
Postbus 20014
2500 EA S-GRAVENHAGE. Tel. nr. 070.789911
Voor bestellingen: Tel. 070.789208
NEW ZEALAND – NOUVELLE-ZÉLANDE
Publications Section,
Government Printing Office Bookshops:
AUCKLAND: Retail Bookshop: 25 Rutland Street,
Mail Orders: 85 Beach Road, Private Bag C.P.O.
HAMILTON: Retail Ward Street,
Mail Orders, P.O. Box 857
WELLINGTON: Retail: Mulgrave Street (Head Office),
Cubacade World Trade Centre
Mail Orders: Private Bag
CHRISTCHURCH: Retail: 159 Hereford Street,
Mail Orders: Private Bag
DUNEDIN: Retail: Princes Street
Mail Order: P.O. Box 1104
NORWAY – NORVÈGE
J.G. TANUM A/S Karl Johansgate 43
P.O. Box 1177 Sentrum OSLO 1. Tel. (02) 80.12.60
PAKISTAN
Mirza Book Agency, 65 Shahrah Quaid-E-Azam, LAHORE 3.
Tel. 66839
PHILIPPINES
National Book Store, Inc.
Library Services Division, P.O. Box 1934, MANILA.
Tel. Nos. 49.43.06 to 09, 40.53.45, 49.45.12
PORTUGAL
Livraria Portugal, Rua do Carmo 70-74,
1117 LISBOA CODEX. Tel. 360582/3
SPAIN – ESPAGNE
Mundi-Prensa Libros, S.A.
Castelló 37, Apartado 1223, MADRID-1. Tel. 275.46.55
Libreria Bosch, Ronda Universidad 11, BARCELONA 7.
Tel. 317.53.08, 317.53.58
SWEDEN – SUÈDE
AB CE Fritzes Kungl Hovbokhandel,
Box 16 356, S 103 27 STH, Regeringsgatan 12,
DS STOCKHOLM. Tel. 08/23.89.00
SWITZERLAND – SUISSE
OECD Publications and Information Center
4 Simrockstrasse 5300 BONN. Tel. (0228) 21.60.45
Local Agents/Agents locaux
Librairie Payot, 6 rue Grenus, 1211 GENÈVE 11. Tel. 022.31.89.50
TAIWAN – FORMOSE
Good Faith Worldwide Int'l Co., Ltd.
9th floor, No. 118, Sec. 2
Chung Hsiao E. Road
TAIPEI. Tel. 391.7396/391.7397
THAILAND – THAILANDE
Suksit Siam Co., Ltd., 1715 Rama IV Rd,
Samyan, BANGKOK 5. Tel. 2511630
TURKEY – TURQUIE
Kültur Yayinlari Is-Türk Ltd. Sti.
Atatürk Bulvari No : 77/B
KIZILAY/ANKARA. Tel. 17 02 66
Dolmabahce Cad. No : 29
BESIKTAS/ISTANBUL. Tel. 60 71 88
UNITED KINGDOM – ROYAUME-UNI
H.M. Stationery Office, P.O.B. 569,
LONDON SE1   9NH. Tel. 01.928.6977, Ext. 410 or
49 High Holborn, LONDON WC1V 6 HB (personal callers)
Branches at: EDINBURGH, BIRMINGHAM, BRISTOL,
MANCHESTER, BELFAST.
UNITED STATES OF AMERICA – ÉTATS-UNIS
OECD Publications and Information Center, Suite 1207,
1750 Pennsylvania Ave., N.W. WASHINGTON, D.C.20006 – 4582
Tel. (202) 724.1857
VENEZUELA
Libreria del Este, Avda. F. Miranda 52, Edificio Galipan,
CARACAS 106. Tel. 32.23.01/33.26.04/33.24.73
YUGOSLAVIA – YOUGOSLAVIE
Jugoslovenska Knjiga, Terazije 27, P.O.B. 36, BEOGRAD.
Tel. 621.992

Les commandes provenant de pays où l'OCDE n'a pas encore désigné de dépositaire peuvent être adressées à :
OCDE, Bureau des Publications, 2, rue André-Pascal, 75775 PARIS CEDEX 16.

Orders and inquiries from countries where sales agents have not yet been appointed may be sent to:
OECD, Publications Office, 2 rue André-Pascal, 75775 PARIS CEDEX 16.   65716-10-1982

OECD PUBLICATIONS, 2, rue André-Pascal, 75775 PARIS CEDEX 16 - No. 42391 1982
PRINTED IN FRANCE
(97 82 08 1) ISBN 92-64-12386-5